YA Non Fiction
YA 530 K635p

Kirkland, Kyle. Physical sciences : notable
research and discoveries
9001028345

DISCARDED
BY THE
MEAD PUBLIC LIBRARY
SHEBOYGAN, WISCONSIN

D0206152

FRONTIERS OF SCIENCE

PHYSICAL SCIENCES

FRONTIERS OF SCIENCE

PHYSICAL SCIENCES

Notable Research and Discoveries

KYLE KIRKLAND, PH.D.

Facts On File
An imprint of Infobase Publishing

PHYSICAL SCIENCES: Notable Research and Discoveries

Copyright © 2010 by Kyle Kirkland, Ph.D.

All rights reserved. No part of this book may be reproduced or utilized in any form or by any means, electronic or mechanical, including photocopying, recording, or by any information storage or retrieval systems, without permission in writing from the publisher. For information contact:

Facts On File, Inc.
An imprint of Infobase Publishing
132 West 31st Street
New York NY 10001

Library of Congress Cataloging-in-Publication Data

Kirkland, Kyle.
   Physical sciences : notable research and discoveries / Kyle Kirkland.
     p. cm. — (Frontiers of science)
   Includes bibliographical references and index.
   ISBN 978-0-8160-7444-0
   1. Physics. I. Title.

   QC21.3.K536 2010
   530—dc22   2009027496

Facts On File books are available at special discounts when purchased in bulk quantities for businesses, associations, institutions, or sales promotions. Please call our Special Sales Department in New York at (212) 967-8800 or (800) 322-8755.

You can find Facts On File on the World Wide Web at http://www.factsonfile.com

Excerpts included herewith have been reprinted by permission of the copyright holders; the author has made every effort to contact copyright holders. The publishers will be glad to rectify, in future editions, any errors or omissions brought to their notice.

Text design by Kerry Casey
Composition by Mary Susan Ryan-Flynn
Illustrations by Melissa Ericksen and Facts On File
Photo research by Tobi Zausner, Ph.D.
Cover printed by Bang Printing, Inc., Brainerd, Minn.
Book printed and bound by Bang Printing, Inc., Brainerd, Minn.
Date printed: June 2010
Printed in the United States of America

10 9 8 7 6 5 4 3 2 1

This book is printed on acid-free paper.

9001028345

CONTENTS

PREFACE

Discovering what lies behind a hill or beyond a neighborhood can be as simple as taking a short walk. But curiosity and the urge to make new discoveries usually require people to undertake journeys much more adventuresome than a short walk, and scientists often study realms far removed from everyday observation—sometimes even beyond the present means of travel or vision. Polish astronomer Nicolaus Copernicus's (1473–1543) heliocentric (Sun-centered) model of the solar system, published in 1543, ushered in the modern age of astronomy more than 400 years before the first rocket escaped Earth's gravity. Scientists today probe the tiny domain of atoms, pilot submersibles into marine trenches far beneath the waves, and analyze processes occurring deep within stars.

Many of the newest areas of scientific research involve objects or places that are not easily accessible, if at all. These objects may be trillions of miles away, such as the newly discovered planetary systems, or they may be as close as inside a person's head; the brain, a delicate organ encased and protected by the skull, has frustrated many of the best efforts of biologists until recently. The subject of interest may not be at a vast distance or concealed by a protective covering, but instead it may be removed in terms of time. For example, people need to learn about the evolution of Earth's weather and climate in order to understand the changes taking place today, yet no one can revisit the past.

Frontiers of Science is an eight-volume set that explores topics at the forefront of research in the following sciences:

- biological sciences
- chemistry

- computer science
- Earth science
- marine science
- physics
- space and astronomy
- weather and climate

The set focuses on the methods and imagination of people who are pushing the boundaries of science by investigating subjects that are not readily observable or are otherwise cloaked in mystery. Each volume includes six topics, one per chapter, and each chapter has the same format and structure. The chapter provides a chronology of the topic and establishes its scientific and social relevance, discusses the critical questions and the research techniques designed to answer these questions, describes what scientists have learned and may learn in the future, highlights the technological applications of this knowledge, and makes recommendations for further reading. The topics cover a broad spectrum of the science, from issues that are making headlines to ones that are not as yet well known. Each chapter can be read independently; some overlap among chapters of the same volume is unavoidable, so a small amount of repetition is necessary for each chapter to stand alone. But the repetition is minimal, and cross-references are used as appropriate.

Scientific inquiry demands a number of skills. The National Committee on Science Education Standards and Assessment and the National Research Council, in addition to other organizations such as the National Science Teachers Association, have stressed the training and development of these skills. Science students must learn how to raise important questions, design the tools or experiments necessary to answer these questions, apply models in explaining the results and revise the model as needed, be alert to alternative explanations, and construct and analyze arguments for and against competing models.

Progress in science often involves deciding which competing theory, model, or viewpoint provides the best explanation. For example, a major issue in biology for many decades was determining if the brain functions as a whole (the holistic model) or if parts of the brain carry out specialized functions (functional localization). Recent developments in brain imaging resolved part of this issue in favor of functional localization by showing that specific regions of the brain are more active during certain tasks. At the same time, however, these experiments have raised other questions that future research must answer.

The logic and precision of science are elegant, but applying scientific skills can be daunting at first. The goals of the Frontiers of Science set are to explain how scientists tackle difficult research issues and to describe recent advances made in these fields. Understanding the science behind the advances is critical because sometimes new knowledge and theories seem unbelievable until the underlying methods become clear. Consider the following examples. Some scientists have claimed that the last few years are the warmest in the past 500 or even 1,000 years, but reliable temperature records date only from about 1850. Geologists talk of volcano hot spots and plumes of abnormally hot rock rising through deep channels, although no one has drilled more than a few miles below the surface. Teams of neuroscientists—scientists who study the brain—display images of the activity of the brain as a person dreams, yet the subject's skull has not been breached. Scientists often debate the validity of new experiments and theories, and a proper evaluation requires an understanding of the reasoning and technology that support or refute the arguments.

Curiosity about how scientists came to know what they do—and why they are convinced that their beliefs are true—has always motivated me to study not just the facts and theories but also the reasons why these are true (or at least believed). I could never accept unsupported statements or confine my attention to one scientific discipline. When I was young, I learned many things from my father, a physicist who specialized in engineering mechanics, and my mother, a mathematician and computer systems analyst. And from an archaeologist who lived down the street, I learned one of the reasons why people believe Earth has evolved and changed—he took me to a field where we found marine fossils such as shark's teeth, which backed his claim that this area had once been under water! After studying electronics while I was in the air force, I attended college, switching my major a number of times until becoming captivated with a subject that was itself a melding of two disciplines—biological psychology. I went on to earn a doctorate in neuroscience, studying under physicists, computer scientists, chemists, anatomists, geneticists, physiologists, and mathematicians. My broad interests and background have served me well as a science writer, giving me the confidence, or perhaps I should say chutzpah, to write a set of books on such a vast array of topics.

Seekers of knowledge satisfy their curiosity about how the world and its organisms work, but the applications of science are not limited to intellectual achievement. The topics in Frontiers of Science affect society on a multitude of levels. Civilization has always faced an uphill battle to procure scarce resources, solve technical problems, and maintain order. In modern times, one of the most important resources is energy, and the physics of fusion potentially offers a nearly boundless supply. Technology makes life easier and solves many of today's problems, and nanotechnology may extend the range of devices into extremely small sizes. Protecting one's personal information in transactions conducted via the Internet is a crucial application of computer science.

But the scope of science today is so vast that no set of eight volumes can hope to cover all of the frontiers. The chapters in Frontiers of Science span a broad range of each science but could not possibly be exhaustive. Selectivity was painful (and editorially enforced) but necessary, and in my opinion, the choices are diverse and reflect current trends. The same is true for the subjects within each chapter—a lot of fascinating research did not get mentioned, not because it is unimportant, but because there was no room to do it justice.

Extending the limits of knowledge relies on basic science skills as well as ingenuity in asking and answering the right questions. The 48 topics discussed in these books are not straightforward laboratory exercises but complex, gritty research problems at the frontiers of science. Exploring uncharted territory presents exceptional challenges but also offers equally impressive rewards, whether the motivation is to solve a practical problem or to gain a better understanding of human nature. If this set encourages some of its readers to plunge into a scientific frontier and conquer a few of its unknowns, the books will be worth all the effort required to produce them.

ACKNOWLEDGMENTS

Thanks go to Frank K. Darmstadt, executive editor at Facts On File, and the rest of the staff for all their hard work, which I admit I sometimes made a little bit harder. Thanks also to Tobi Zausner for researching and locating so many great photographs. I also appreciate the time and effort of a large number of researchers who were kind enough to pass along a research paper or help me track down some information.

INTRODUCTION

In 1687, the British physicist Sir Isaac Newton (1642–1727) made a startling announcement—the force that makes an apple fall to the ground is the same force that keeps planets in their orbits. Newton's discovery of the law of universal gravitation unified many observations on Earth as well as in space. Some of the most impressive advances in science occur when a theory or equation explains a wide range of phenomena in one elegant statement or formula.

But as researchers probe further into the frontiers of science, unexpected findings often turn up. Even the most elegant theory can get called into question. While Newton's universal law of gravitation applies to many situations and remains an important and frequently used theory, the German-American physicist Albert Einstein (1879–1955) studied its weaknesses, such as its inability to account for all of the precession in Mercury's perihelion (the point in its orbit at which the planet is closest to the Sun—this point slowly moves, or precesses, after each revolution). In 1916, Einstein formulated the *general theory of relativity,* which is a more comprehensive and accurate theory of gravitation.

Physical Sciences, one volume in the Frontiers of Science set, is devoted to researchers who expand the frontiers of physics—and often uncover phenomena that contradict prevailing wisdom. Physics is the study of matter and energy and how objects move and change. The term *physics* derives from a Greek word *physikos,* which means "of nature." Physics is the study of nature in its essential forms, and its goal is to explain as much of the world as possible in the most concise and accurate manner, as the ancient Greeks attempted in theories such as the four fundamental substances—earth, air, water, and fire—that they believed comprised the universe. In addition to the intellectual satisfaction of understanding how nature works, advances in physics offer tremendous benefits such as cleaner, cheaper energy sources. People have pursued physics knowledge for a long time, but while physics is a mature science, it is by no means finished, as this book will show.

This book discusses six main topics, each of which comprises a chapter that explores one of the frontiers of physics. Reports published in journals, presented at conferences, and issued in news releases describe research problems of interest in physics, and how scientists are tackling these problems. This book discusses a selection of these reports—unfortunately there is room for only a fraction of them—that offers students and other readers insights into the methods and applications of physics.

Physics can be a complicated subject, especially at the frontiers. Students and other readers need to keep up with the latest developments, but they have difficulty finding a source that explains the basic concepts while discussing the background and context that is essential to see the big picture. This book describes the evolution of ideas and explains the problems that researchers are presently investigating and the methods they are developing to solve them. No special mathematical knowledge is required to understand the material presented in this volume.

Chapter 1 describes fusion, the process in which atomic nuclei join and release enormous amounts of energy. People began building nuclear weapons based on fusion in the 1950s, but physicists have been unable to develop an economical method of using controlled fusion reactions to generate electricity and other useful forms of energy. Fusion is a highly desirable energy source because it releases little pollution and its fuel is cheap and abundant. Several ongoing projects aim to create an economical power source based on fusion, and if they are successful, the energy demands of the world can be met in an environmentally friendly way.

The study of atoms and their components involves large amounts of energy per particle. To create the necessary conditions, physicists employ giant machines called particle accelerators, the subject of chapter 2. The electric and magnetic fields of these machines boost particles up to nearly the speed of light and send them hurtling into one another in violent collisions. Physicists study the debris of these collisions to learn more about the fundamental nature of particles, which are not composed of the four substances that early philosophers imagined, but can be classified in other important ways. Particle physics also provides valuable clues on the nature of the universe—perhaps a surprising result from the study of such small objects.

Scientists have recently focused their attention on one specific class of particle—neutrinos, the subject of chapter 3. These mysterious particles blithely zip through stars and planets, rarely stopping to interact with other pieces of matter. Neutrino properties such as mass, which has yet to be quantified, are essential aspects of particle physics, but even gifted (and well-funded) researchers have difficulty studying a particle that hardly interacts with anything. Physicists have been forced to develop novel methods to measure these elusive and ghostly particles.

Chapter 4 describes the most efficient means of electrical conduction—superconductors. Electricity is a critical component of many technologies, including the particle accelerators of chapter 2, but ordinary conductors resist the flow of current, introducing serious losses and limiting the usefulness of electrical equipment. Superconductors have no resistance. Set up a current in a superconductor, and it will keep going forever! Most superconductors require extremely low temperatures to function, but researchers have recently found several classes of material that can operate at higher temperatures. No one fully understands how these new superconductors work, however, and a comprehensive theory to guide future research is one of the major goals of modern physics.

Since physics deals with fundamental subjects, other branches of science often employ the methods and principles of physics. Such is the case for the study of how complex objects or systems of objects evolve. Researchers from a variety of disciplines, including scientists who study storm systems and those who study brain systems, have found surprising patterns in the behavior of complex systems. These findings heralded chaos theory, as discussed in chapter 5. Order and predictability sometimes arise out of seemingly chaotic and random phenomena. Scientists are studying the patterns to learn more about complicated systems such as weather, the brain, and atomic interactions.

One of the most fundamental questions concerns the nature of matter. Although particle physicists have peered into the very heart of matter, no one is certain what matter is ultimately made of—or even if there is an answer to this question. Chapter 6 deals with a theory called string theory, which posits that matter consists of thin, vibrating strings. This theory is elegant but mathematically complex, and researchers have yet to find experiments with which they can test the ideas. Scientific theories are of little use without tests that support (or reject) them. Physicists who delve into the foundations of physics are actively pursuing some kind of experiment that will serve as a reality check for the fascinating but speculative ideas of string theory.

These topics offer a sample of the frontiers of physics. Many surprising discoveries have recently come to light, some of which have been explained, and some of which have not. Explanations for those that remain mysterious, as well as more and possibly greater discoveries, await the insight of future researchers.

Nuclear Fusion: Power from the Atom

In 1938, the German-American physicist Hans Bethe (1906–2005) discovered how nuclear fusion powers the Sun and other stars. According to an old story, before Bethe published his discovery he was walking late at night with his fiancée, Rose. While Rose gazed at the bright stars, Hans bragged that he was the only person in the world who knew how they shine. One might perhaps view this story as something of a legend or myth, and the physicist Ralph Wijers, who visited Bethe's house in 1999, asked about it. In an article published in a 2007 issue of the *Bulletin of the American Astronomical Society,* Wijers wrote, "Hans grinned a bit sheepishly, but Rose roundly confirmed the story with a big smile. Not too impressed, she had replied: 'That's nice.' And so it was."

Fusion is a nuclear reaction in which atomic nuclei (plural of *nucleus*) join or fuse. The process liberates an enormous quantity of energy. This energy is sufficient to keep the Sun and other stars shining brightly for a long time and can also make a frighteningly destructive bomb.

Although the study of nuclear fusion has taught researchers much about the physics of atoms and nuclei, the seven decades since Bethe's discovery have been disappointing in at least one major respect. If scientists and engineers could learn how to control fusion in a reliable and safe manner, it would solve the world's energy problems. A solution is badly needed: Fossil fuels such as oil presently supply most of the world's energy, but these fuels are rapidly being depleted and their combustion pollutes the environment and releases *greenhouse gases.* These gases trap heat, warming the Earth's surface and melting glacial ice, leading to deleterious effects such as rapidly rising sea levels. In a 2007 report entitled "Climate Change 2007," the Intergovernmental Panel on Climate Change (IPCC), a scientific organization established by the United Nations, concluded, "Continued greenhouse gas emissions at or above current rates would cause further warming and induce many changes in the global climate system during the 21st century that would very likely be larger than those observed during the 20th century." Fusion power would produce little *radioactive* waste or greenhouse gases, and the necessary materials are abundant and inexpensive. But as yet, no one has found a way to design a viable fusion reactor or power plant.

Decades of research on nuclear fusion have generated substantial progress as well as a considerable amount of controversy. Controversy should be expected in a research field that, if successful, offers an almost boundless supply of cheap, environmentally friendly energy. But the controversies and disappointments over the years have taken a toll, and other approaches to alternative energy, such as fuel cells and solar energy, tend to get more attention these days. The journalist Dan Clery wrote in the October 13, 2006, issue of *Science* that "skeptics joke that 'Fusion is the power of the future and always will be.'" But researchers at the frontiers of physics are soldiering onward, and some are having considerable success. This chapter explains the basic concepts of fusion, discusses the controversy of *cold fusion,* and describes projects in which people have invested a lot of time and money to build a viable fusion power plant.

INTRODUCTION

Before Bethe, scientists had only a hazy idea of what keeps the Sun shining. Early researchers knew that if the fuel were coal or oil or some other combustible material familiar to 19th-century physicists the Sun would not last long. The German scientist Hermann Helmholtz (1821–94) and the Scottish physicist Sir William Thompson, Lord Kelvin (1824–1907), theorized that gravitational energy powered the Sun. According to this theory, the Sun's great mass contracts under the force of gravitation. To see how this might work, think of gravitational potential energy, such as that of a rock poised on top of a cliff or the raised weights of a grandfather clock. When the weight falls, its potential energy (due to its height) gets converted into kinetic energy—the energy of motion—and, in a grandfather clock, some of this energy is used to swing the pendulum. In the Kelvin-Helmholtz theory, the energy of the falling surface of the Sun gets converted into heat and radiation.

Kelvin calculated that gravitational energy could power a body the size of the Sun for about 20 or 30 million years. The British naturalist Charles Darwin (1809–82) found this troubling because he believed his theory of evolution required a much longer time over which to act. Later, scientists discovered the age of the Sun and solar system is about 4.5 billion years old, much older than Kelvin's calculation. Although astronomers now believe the Kelvin-Helmholtz theory does hold true in certain cases, the Sun's source of energy lies elsewhere.

An important clue came in 1896. In the course of some experiments, the French physicist Henri Becquerel (1852–1908) discovered radioactivity—the emission of energetic particles or radiation by certain elements, in this case uranium. A few years later, the Polish scientist Marie Curie (1867–1934) and her husband, the French researcher Pierre Curie (1859–1906), found other radioactive elements and characterized their properties. The energy was coming from reactions of the atom's nucleus, the central portion of the atom that the New Zealand-British physicist Ernest Rutherford (1871–1937) and his colleagues discovered with a set of experiments conducted in the early 20th century.

An atom is composed of negatively charged electrons swarming around a tightly compacted nucleus of positively charged protons and electrically neutral neutrons, as shown in the figure at the top of page 4. All atoms of the same element have the same number of protons in the nucleus—this number, the atomic number, identifies the element. All carbon atoms have six protons, for example, and hydrogen atoms have one. But the number of neutrons can vary among atoms of the same element. *Isotopes* are atoms that have the same number of protons but a different number of neutrons. For example, the most common form of hydrogen has one proton and no neutrons, and is represented by the symbol ^1H. (The number at the upper left stands for the number of *nucleons*—protons and neutrons.) *Deuterium,* ^2H, with one proton and one neutron, and *tritium,* ^3H, with one proton and two neutrons, are isotopes of hydrogen.

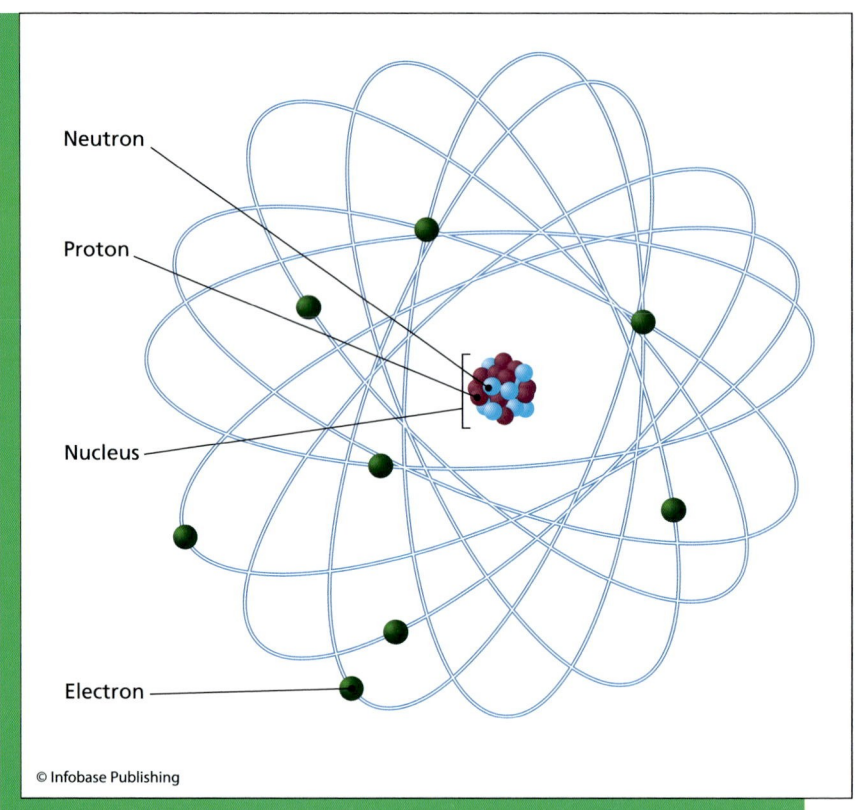

Neutron

Proton

Nucleus

Electron

© Infobase Publishing

A simple model of the atom consists of protons and neutrons in the nucleus, surrounded by "orbiting" electrons.

The compactness of the nucleus would appear electrically impossible, since protons repel each other (positive charges repel other positive charges and attract negative ones). What accounts for the ability of nuclear protons to overcome this repulsion is the existence of a force called the *strong nuclear force,* or just *strong force.* The strong force acts only over extremely short distances. Electrical repulsion normally keeps protons away from each other, but when protons find themselves in close quarters—which might happen, for instance, if two high-speed protons collide—the strong force takes over, gluing the particles together with enough strength to withstand the electrical force that keeps trying to pull them apart.

Electrons are involved in a lot of phenomena, such as forming bonds during chemical reactions, but this chapter focuses on the atom's nucleus. The nuclei of some atoms are naturally unstable and, as Becquerel and the Curies discovered, spontaneously decay into other forms, emitting certain particles or radiation in the process. For example, tritium (^{3}H) nuclei are unstable, and in a little more than 12 years half of a group of tritium nuclei undergoes a process known as *beta decay,* whereby one of the neutrons becomes a proton, and the nucleus emits particles (one of which is an electron, generated during the decay). The tritium nucleus becomes an isotope of helium, ^{3}He, with two protons in the nucleus and one neutron.

In addition to spontaneous radioactive decay, nuclear reactions occur when particles collide or get absorbed into a nucleus. The two basic types of reactions are fission and fusion. Fission occurs when a nucleus splits, or fissions. This is the reaction that powered the earliest atomic weapons, such as the two bombs dropped on Japan to end World War II in 1945. Fission is also the process by which all *nuclear reactors* in

This view of the nuclear reactor at Dungeness B nuclear power station in Kent, England, shows the top of the plate in which uranium fuel rods are housed. *(Jerry Mason/Photo Researchers, Inc.)*

Einstein's Famous Equation: $E = mc^2$

As a 26-year-old patent officer in Bern, Switzerland, Albert Einstein published several papers that helped establish the foundations of modern physics. These publications involved the special theory of relativity and quantum physics, which amended the classical laws of physics as formulated by Sir Isaac Newton. One paper, submitted to the German journal *Annalen der Physik* (*Annals of Physics*) in 1905, was a short, three-page article whose title in English was, "Does the Inertia of a Body Depend on Its Energy Content?" Using some of the ideas he had published earlier on relative motion and the speed of light, c (which Einstein correctly postulated is constant), Einstein answered the title's question in the affirmative: Mass, m (inertia), is related to energy, E, by the equation $E = mc^2$.

Since the speed of light is constant, the formula says that energy is proportional to mass, and the constant of proportionality, c^2, is huge. The speed of light is about 186,000 miles/sec (300,000 km/sec) in a vacuum. Squaring this large number makes it even more enormous. Thanks to the magnitude of c^2, a little mass goes a long way as far as energy is concerned.

No one paid too much attention to this equation until researchers began to understand nuclear processes such as radioactive decay, fission, and fusion. After Hahn and Strassman discovered fission of uranium nuclei in 1938, researchers began thinking about harnessing this enormous quantity of nuclear energy. The Hungarian physicists Leó Szilárd and Eugene Wigner worried that oppressive governments such as the Nazis would develop fearsome nuclear weapons. They wanted to warn the Americans of the danger, but they were worried their concerns would go unnoticed. In order to maximize the

A cloud of smoke and debris rises 20,000 feet (6,100 m) above Hiroshima, Japan, on August 6, 1945, after a U.S. bomber drops an atomic bomb. *(U.S. National Archives)*

impact of their warning, they decided to enlist one of the most famous scientists of all time—Albert Einstein. Einstein, who had fled Germany after the Nazis gained power in 1933, agreed to help raise the alarm. His 1939 letter to President Franklin D. Roosevelt got the political leader's attention, and the United States went on to develop an atomic bomb. As it turned out, the United States was the only country to succeed in developing an atomic bomb during World War II. Two atomic bombs dropped on Japan ended the war in 1945.

Other than the letter to Roosevelt, Einstein had little to do with the development of the bomb. As a pacifist, Einstein generally opposed military activities. But Walter Isaacson, in his 2007 biography, *Einstein,* wrote: "Between the influence imputed to that letter and the underlying relationship between energy and mass that he had formulated forty years earlier, Einstein became associated in the popular imagination with the making of the atom bomb, even though his involvement was marginal." It was not an association that Einstein was proud of. In 1947, Einstein remarked, "Had I known that the Germans would not succeed in producing an atomic bomb, I never would have lifted a finger."

operation today generate their power. A typical fission reaction occurs when a nucleus such as the uranium isotope ^{235}U absorbs a neutron, which might cause it to split into two lighter fragments—a barium isotope, ^{141}Ba, and a krypton isotope, ^{92}Kr—and release a few neutrons at the same time. The German scientists Otto Hahn (1879–1968) and Fritz Strassman (1902–80) were the first researchers to observe a fission reaction in uranium in 1938.

Fusion is the opposite of fission. In a fusion reaction, two lighter nuclei join to form a larger nucleus. For example, ^{1}H and ^{2}H may combine to form ^{3}H, or ^{1}H and ^{3}He may form ^{4}He.

Both fission and fusion reactions release a prodigious amount of energy. The reason for this is that the nucleons of a nucleus are bound tightly, and the energy of this bond is known as the binding energy. Albert Einstein formulated a simple equation in 1905—long before anyone knew of its application to chemistry and nuclear physics—that equates this energy, E, to the product of the mass, m, and the square of the speed of light in a vacuum, c. In a fission or fusion reaction, for example, the products have slightly less mass than the reactants. This mass gets transformed into energy according to Einstein's equation, $E = mc^2$, as described in the sidebar on page 6.

One nucleus by itself has little energy, but in a chain reaction, which occurs in nuclear weapons, the total energy liberated is enough to create an intense fireball of heat and radiation. When the process is carefully controlled, fission can also safely release enough energy to drive huge electric generators. In 2008, nuclear reactors—all of which today are based on fission—produce about 15 percent of the world's electricity, including 19.7 percent of electricity in the United States, according to the World Nuclear Association (WNA).

But the problems with fission reactors are severe. The common fuel, uranium, is found in nature in a mixture of isotopes, only one of which, ^{235}U, efficiently enters into fission reactions. This isotope comprises only a few percent of uranium; in order to be of any use to reactors, uranium commonly needs to be enriched, raising the proportion of ^{235}U, which is usually an expensive procedure. And since uranium is a rare resource that is being quickly depleted, there is a danger of running out of fuel in the future.

Another serious problem with fission is that the spent fuel continues to be highly radioactive and quite dangerous, since the levels of emissions are sufficient to cause radiation damage and long-term diseases such as cancer. Storing this radioactive waste safely is costly, requiring strong containers to prevent spillage and a place to keep them. People in the vicinity of these storage places are not usually very happy about it.

THE POWER OF FUSION

Fusion avoids most of the problems of fission. There are few dangerous or environmentally hazardous emissions, and the fuel is abundant and cheap.

The same principles of nuclear physics apply to fusion reactions. In a fusion reaction, a small amount of mass gets converted into energy. For example, when deuterium fuses with tritium, the products have about 0.3 percent less mass—this is the mass that gets transformed into energy, by Einstein's formula $E = mc^2$. Although the percentage seems a trifling amount, the magnitude of the c^2 term assures that this process generates a lot of energy. Fusion is slightly more efficient than uranium fission, because uranium fission reactions generally convert only about 0.1 percent of their mass into energy.

The Australian physicist Sir Mark Oliphant (1901–2000) and his colleagues observed fusion reactions in hydrogen nuclei in 1932, although the details of the process and its role in powering the Sun were not known until Bethe's calculations a few years later. In 1952, the United States tested the first H-bomb—hydrogen bomb, a nuclear weapon employing the fusion of hydrogen nuclei. The former Union of Soviet Socialist Republics (USSR) tested a hydrogen bomb in 1953.

Fusion weapons took a little longer to construct than fission bombs because of the extreme conditions required for fusion to occur. Since nuclei are positively charged, their electrical forces repel one another, so nuclei are not normally found close together. But certain conditions overcome this electrical repulsion. Higher temperatures correspond with greater movement of atoms and molecules—the reason heat causes materials such as steel bridges to expand is that the volume increases due to this greater motion. Exceptionally high temperatures cause electrons to fly away from their atoms and nuclei to crash together. High pressures also reduce nuclei distances, since the pressure squeezes particles together.

Such extremes in temperature and pressure occur in large objects such as the Sun. The Sun consists of mostly hydrogen and helium gases and has a radius of about 434,000 miles (700,000 km), with a mass more than 300,000 times larger than Earth. No one is certain of the temperature of the Sun's core, but scientists believe it can be as hot as 27,000,000°F (15,000,000°C).

How do researchers study the Sun's interior? In addition to theoretical calculations, astronomers can observe certain particles coming from the Sun. The most important particles are neutrinos, the subject of chapter 3 of this book. Scientists detect these particles and use their knowledge of nuclear reactions to study fusion processes occurring in the Sun.

The extreme conditions inside the Sun provide an unmistakable hint as to why the technological development of fusion has been slower than fission. High temperature and pressure are normal in the Sun's core, but replicating such conditions on Earth's surface is enormously costly. Generating these conditions for a brief instant, such as in a bomb, is not so hard, but a power-generating reactor must involve slow, controlled reactions. In order for any kind of generator to be economical, it must produce more power than it consumes. This problem lies at the heart of the trouble that has plagued fusion power research for the last 50 years.

There is a possibility that such extreme conditions are not actually essential for fusion to occur. In other words, certain kinds of fusion events may happen even in much milder environments. This possibility, including cold fusion, is controversial and will be discussed in the final two sections of the chapter. Many researchers are convinced that fusion generally requires extreme conditions and have set about reproducing these conditions in the laboratory.

INERTIAL CONFINEMENT—IGNITION WITH LASERS

The material in the Sun is called *plasma.* High temperatures strip the electrons from atoms, producing electrical charges called *ions.* Plasma is a state of matter consisting of ions in the gaseous state. This state of matter does not behave the same way as an ordinary gas because of the electrical charges. For instance, a plasma responds to electric and magnetic fields.

To create the conditions under which fusion typically occurs, researchers need to heat a plasma to millions of degrees. Keeping this exceptionally hot material confined so that the nuclei can undergo fusion is a big problem. In the interior of the Sun, the enormous gravitational forces exert enough pressure to keep the nuclei confined tightly enough

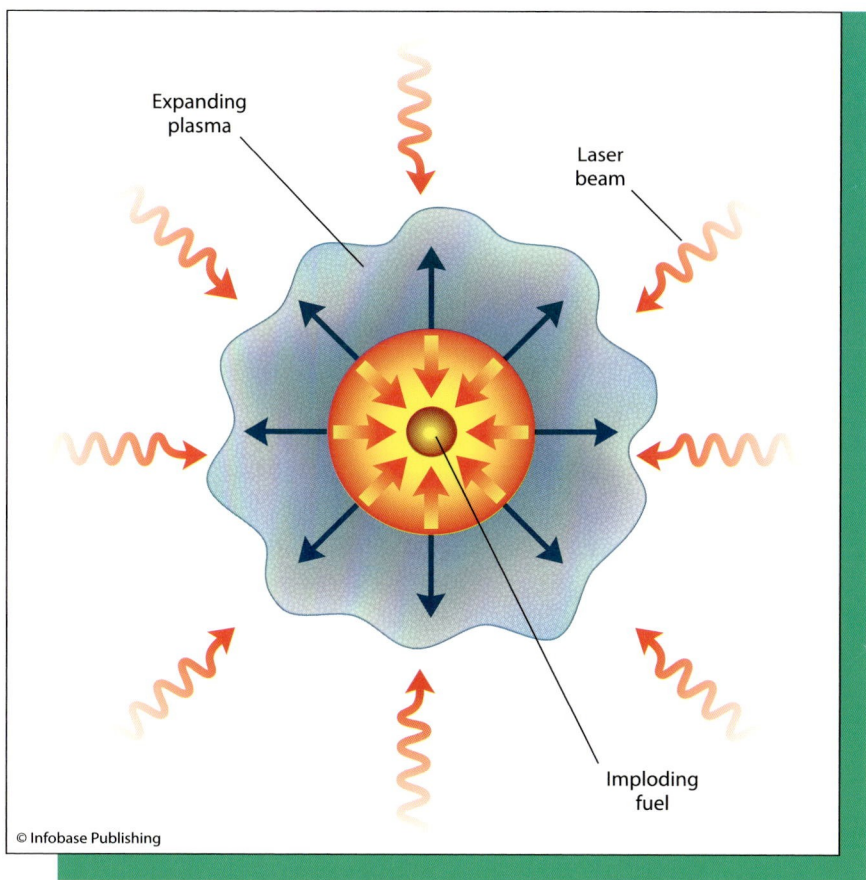

Expanding plasma

Laser beam

Imploding fuel

© Infobase Publishing

Lasers heat a fuel pellet, causing the interior to implode.

for fusion to occur. On the surface of Earth, the usual means of confining a material is to use some sort of container, but confining a material having a temperature of millions of degrees presents a variety of difficulties. The container must be able to withstand such temperatures without melting, and, just as important, the walls of the container should not cool the material to such an extent that fusion events become rare or impossible.

Two main techniques of confinement have been studied. The technique described in this and the following section is called inertial confinement. Inertia is the opposition of a body to a change in its motion—a resting body requires a force, such as a push or pull, to get moving, and a moving body requires a force to slow it down (or change its direction). The idea of inertial confinement is to confine a material for a short period of time by its own inertia. One of the most prominent approaches is to aim a high-energy laser beam at a small pellet of fusable material. Lasers are concentrated sources of light, and a beam with high intensity can deliver a large amount of energy to a small space. As illustrated in the figure at the top of page 11, the laser's energy evaporates the pellet's surface, sending particles flying away. But because of their inertia, the particles cannot move fast enough to keep from blocking the particles in the interior of the pellet, and by Newton's third law—every action has an equal and opposite reaction—the pellet's interior is compressed by a shock wave from the escaping gases. As a result, the pellet's core attains a temperature of millions of degrees and a pressure exceeding that of Earth's atmosphere by millions of times.

Although the technique works, scientists do not fully understand the dynamics of this complex process. Studying this process is complicated because of its speed and extreme conditions, but in 2008 J. R. Rygg at the Massachusetts Institute of Technology (MIT) and his colleagues at that institution and the University of Rochester in New York developed a useful tool. The researchers adapted radiography—the production of images with radiation other than visible light—to take a picture of the activity within the small pellet as it implodes. These pictures revealed previously unobserved electrical and magnetic phenomena occurring in the process, such as an electric field arising from the immense pressure gradient. Knowledge of these fields will be essential to get a better understanding of how inertial confinement works and how to improve it. The researchers published their paper, "Proton Radiography of Inertial Fusion Implosions," in a 2008 issue of *Science*.

To create a facility to study inertial confinement, among other subjects of interest, researchers have built lasers of enormous size and power. One facility, called National Ignition Facility (NIF), contains the world's largest laser system.

NATIONAL IGNITION FACILITY

The goal of NIF is to create "a miniature star on Earth," as their scientists often say. NIF researchers aim to produce events similar to the reactions occurring in stars such as the Sun.

Recreating the conditions inside a star requires concentrating an enormous amount of energy in a small space. NIF has 192 high-power

The National Ignition Facility aims 192 laser beams at a small target area. *(Lawrence Livermore National Security, LLC, and Lawrence Livermore National Laboratory)*

lasers, each aimed at a target the size of a BB pellet. This number of lasers seems a little bit of an overkill—at peak power their beams generate about 1,000 times the electrical generating power of the United States! But the lasers are only switched on for short periods of time, producing exceptionally brief pulses on the order of a nanosecond (one-billionth of a second).

All this energy is needed to produce fusion, and it cannot all come from one laser beam—the beams must deliver the energy symmetrically, the same at each point, so that the pellet is not pushed one way or another. Synchronized delivery means that the lasers must be switched on and aimed with incredible precision. The laser pulses must hit the target within 30 picoseconds—30 trillionths of a second—of one another, and cannot deviate more than about 0.002 inches (0.005 cm). Electrical and optical equipment capable of such precision is sophisticated and extremely expensive.

The *ignition* term in NIF's name comes about when the laser delivers its energy to the target, which consists of hydrogen isotopes such as deuterium. Temperatures rise to millions of degrees and the pressure is equivalent to about 100 billion times that of Earth's atmosphere. Under such conditions, fusion of the hydrogen isotopes can occur.

Housing this enormous laser system is a building the size of a football stadium. The building, located at Livermore, California, and finished in 2001, is 704 feet (214 m) long, 403 feet (123 m) wide, and 85 feet (26 m) tall. NIF is part of the Lawrence Livermore National Laboratory, one of the main government research laboratories in the United States. The Lawrence Livermore National Laboratory, established with the guidance of the University of California, Berkeley, physicist Ernest Lawrence in 1952, has been involved in many large projects, including the development of nuclear fusion bombs and the study of genetic mutations associated with radiation exposure.

Fully operational in 2009, the NIF studies the ignition of fusion in pellets of deuterium and tritium. As fuel for a future nuclear reactor, deuterium is an excellent choice. Deuterium is an extremely common substance—about one out of every 6,000 or 7,000 atoms on Earth is deuterium. A cup of water contains enough deuterium to generate the same amount of energy as 300 times the same quantity of gasoline.

It is important to understand that NIF is a research facility, not a viable reactor. As a reactor it would be terribly uneconomical, since so much energy is required to set up the conditions for fusion to proceed. Consider the enormous power requirements of the 192 lasers. NIF spends more money creating its energy output than it could get from selling this energy to consumers, which means the enterprise would fail from an economic perspective.

As a research facility, however, NIF has great potential. "NIF has been designed to be a platform for cutting-edge science in the decades ahead," said NIF project manager Ed Moses in a 2003 article written by Katie Walter in *Science and Technology Review,* a magazine published by the Lawrence Livermore National Laboratory. The facility will be used to study inertial confinement in the hope that the technique can be improved, both scientifically and economically. NIF will also be instrumental in the study of a broad array of physics topics, including optics and plasma physics.

The hope that a viable fusion option will emerge from the experiments at this facility has no guarantee, despite the expense and sophistication of NIF's equipment, but it is possible. Early computers in the 1940s, for example, were room-sized contraptions full of often-overheating electrical elements, but after a few decades, researchers discovered ways to reduce the cost and size of these machines, tremendously boosting their efficiency.

If inertial confinement should prove to be the most economical approach to fusion power, NIF will have paid enormous dividends. But researchers are not putting all their deuterium atoms in one basket. An alternative technique takes advantage of the electromagnetic properties of plasmas.

MAGNETIC CONFINEMENT—A BOTTLE WITH NO WALLS

Recall that plasmas consist of ions in a gaseous state. Magnetic fields exert a force on a moving electric charge that is perpendicular to its direction of motion. (This force is strongest when the electric charge is traveling perpendicular to the magnetic field's orientation, or lines of force.) In other words, magnetic fields deflect the trajectory of an electric charge by pushing it sideways, at a 90-degree angle to the direction in which it is traveling. If the magnetic field is strong enough, the force deflects the charge's motion so much that the path becomes a circle.

In the magnetic confinement technique, magnetic fields constrain the plasma that is to undergo fusion. This magnetic "bottle" has no physical wall, but uses the force of magnetic fields to deflect any ion that strays too far. There is no force on a stationary charge (or a charge that is moving parallel to the field's lines of force). By a careful positioning of magnets, researchers can confine a plasma without the need for a container that touches the material, which would possibly melt or let too much heat escape.

A sphere would be a good choice in which to sculpt a plasma magnetically, but the required magnetic fields are difficult. As shown in the figure on page 16, most magnetic containers have the shape of a torus—a doughnut shape—although some containers are more spherical. In the most efficient strategy, a current-carrying wire spirals around the doughnut. Electric currents produce magnetic fields, so when charges are flowing through the wire, a magnetic field of a certain orientation surrounds it. Judicious selection of the intensity of the current and the number of coils of the wire produces the desired magnetic confinement. Researchers must also take into account the magnetic fields generated by the moving charges in the plasma.

In the 1950s, scientists in the former USSR developed a device that efficiently exploited magnetic confinement. (One of the developers was Andrei Sakharov [1921–89], a nuclear physicist as well as a human rights activist who protested Russian policies that he believed were oppressive.) The device became known as a tokamak, from an acronym of the Russian words *toroidal'naya kamera s aksial'nym magnitnym polem* (toroidal chamber with axial magnetic field). Electric currents heat the plasma to extremely high temperatures.

Although fusion in the Sun occurs at (only!) 27,000,000°F (15,000,000°C), the heat that drives fusion reactors on Earth needs to be a little more intense, because the plasma is less dense than in the core of a star. Thermonuclear fusion—fusion that is driven by the thermal (heat) motion of the nuclei—needs a temperature of at least 180,000,000°F (100,000,000°C) in order to succeed.

A tokamak is the basis of the Joint European Torus (JET), the largest nuclear fusion research facility in the world at the present time. Located at Culham in the United Kingdom, scientists from all over the European Union use JET to study the tokamak device and thermonuclear fusion. But like NIF, JET is only a step toward understanding the fusion process,

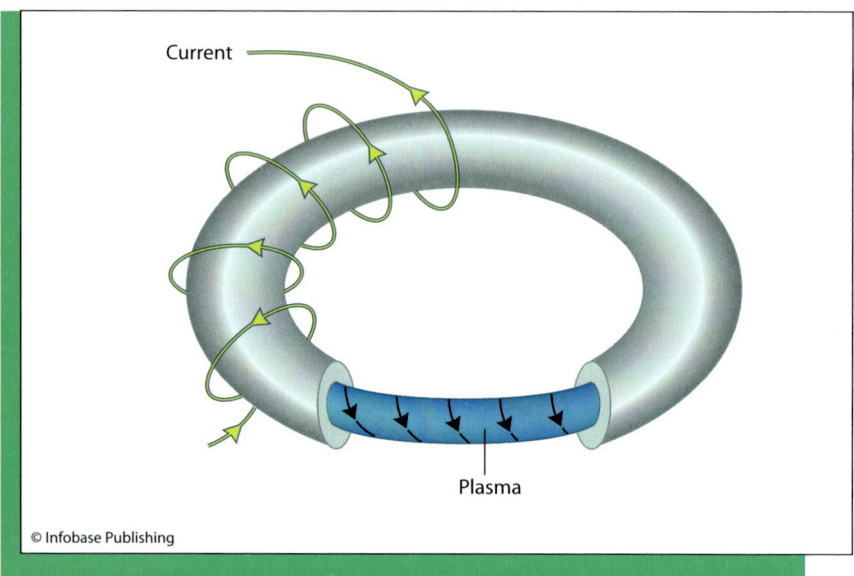

© Infobase Publishing

Magnetic confinement schemes are often toroidal (doughnut-shaped).

Tokamak Fusion Test Reactor located at the Princeton Plasma Physics Laboratory in New Jersey *(U.S. Department of Energy/Photo Researchers, Inc.)*

it is not an economical reactor. JET has met with success in achieving fusion, as described in the sidebar on page 18, but it can only generate about 70 percent of the power it uses to heat the plasma to the required temperature. The device is therefore a consumer rather than a producer.

An alternative magnetic confinement technique is to use the magnetic field of the plasma itself. Charges in motion, whether they consist of a current of electrons in a wire or the flowing ions in a plasma, produce magnetic fields. In a tokamak, magnetic fields generated external to the plasma combine with the plasma magnetic field to confine and control the ions. But in a technique known as the Z-pinch, electrical currents in the plasma create the primary means of magnetic confinement.

A Z-pinch uses the pinch effect: Currents flowing in the same direction generate magnetic fields that create a force that pulls or pinches the charges together. The effect can be demonstrated in a laboratory with two parallel wires, which move closer together when they carry a strong current flowing in the same direction. In a plasma, the ions carry the current, and if the plasma's particles move in concert in the appropriate direction,

Joint European Torus

In 1970, the Council of the European Community decided to invest in fusion power research. (The Council of the European Community evolved into the Council of the European Union, which today is the main policy-making institution of the European Union.) Skyrocketing oil prices in the 1970s— the result of unrest in the Middle East—encouraged this investment, as European politicians and scientists sought additional sources of energy. Design and planning for a fusion device began in 1973, and in 1979 construction started at the selected site, a former Fleet Air Arm airfield at Culham in Oxfordshire, England. Workers finished the job in 1983, and on June 25 of that year, JET scientists initiated the first plasma. Operations today are conducted under the guidance of the European Fusion Development Agreement, which provides the framework for European research into magnetic confinement and thermonuclear fusion.

JET operates the largest tokamak in the world at the present time. (A future project will be larger, as discussed in the following section.) The major radius of the plasma is 9.7 feet (2.96 m) and the minor radius is 6.9 feet (2.1 m). Total volume of the plasma is about 3,180 feet^3 (90 m^3). Several million amps of current are needed to heat this plasma, which is a huge amount of current; powerful car batteries can provide only a few hundred amps.

In 1991, a tritium experiment at JET achieved the first controlled release of fusion power. Later, in 1997, JET produced a world-record 16 megawatts of power from fusion. A megawatt is a unit of power equal to 1 million watts and is a considerably large amount—a typical lightbulb uses 60 watts, and an automobile engine can generate up to a few hundred thousand watts. To produce this fusion power, however, JET required about 24 megawatts of input power to confine and heat the plasma. JET's successes show that fusion power is possible, though at present still not quite economical.

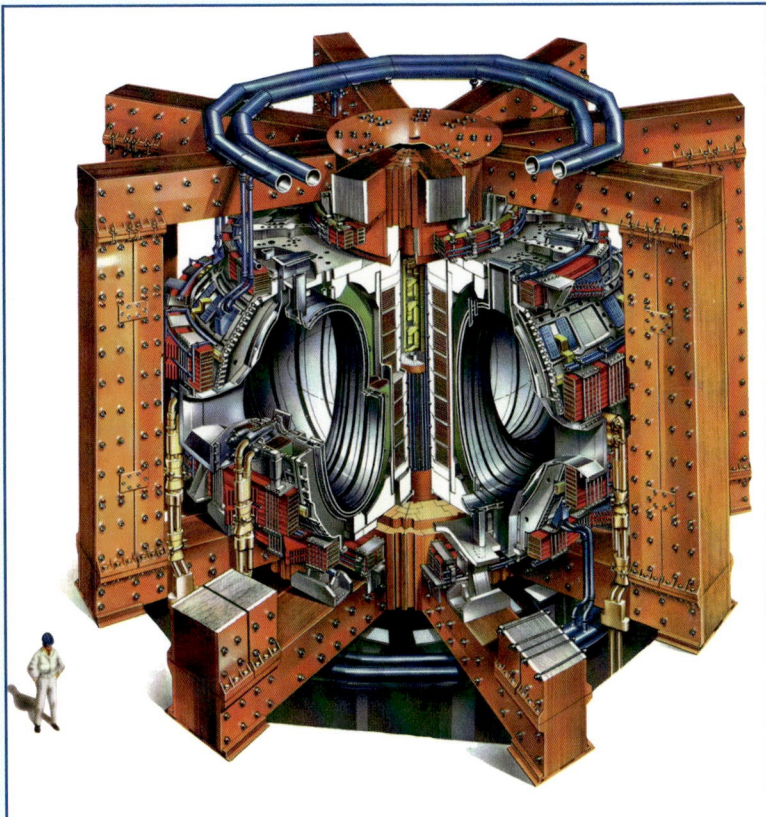

Diagram of the Joint European Torus, with a section cut away to reveal the interior—the person standing at the bottom left of the diagram provides a sense of scale. *(JET, the Joint European Torus)*

the particles experience a magnetic force that squeezes them into a smaller space. The Z term in the Z-pinch's name derives from some of the earliest devices in the 1950s, in which an important component of the magnetic field lay along the Z-axis. (In a three-dimensional coordinate system, the Z-axis is perpendicular to the X- and Y-axes and is usually drawn in the vertical or up-and-down direction.) Researchers continue to study various versions of Z-pinches, although serious problems, such as instabilities that tend to disrupt the process, are formidable obstacles to the potential development of Z-pinch fusion power.

In contrast to Z-pinch devices, JET's success has encouraged scientists and funding agencies to try something even bolder. After about two decades of discussions, on November 21, 2006, the governments of China, India, Japan, South Korea, Russia, the European Union, and the United States formally agreed to support a new and much larger project—ITER.

ITER FUSION

ITER was originally an acronym for the International Thermonuclear Experimental Reactor. People still sometimes use this name, although officials have shied away from the term *thermonuclear* because of the negative connotation of nuclear weapons. Instead, ITER's name is often explained these days in reference to the Latin word *iter,* which means road or way. Supporters of ITER hope that the project paves the way toward the economical use of fusion power.

Some of the thorniest problems in large international projects such as ITER involve the site of the facility—every participant would like to host the facility, but in the case of ITER, only one can do so. In 2005, officials finally reached a consensus to build the reactor in Cadarache in southern France.

A design of ITER has not yet been finalized as of early 2009, but plans call for a tokamak that is much larger than JET. The plasma major radius should be around 20.3 feet (6.2 m) and the minor radius about 6.6 feet (2.0 m), with a volume nearly 10 times that of JET. Researchers expect that fusion power output will reach 500 megawatts. Fuel will consist of the hydrogen isotopes deuterium and tritium.

The enormous facility will not be cheap. Early estimates budgeted about $9 to $12 billion in U.S. currency. But the complexity of the design, as well as future increases in the cost of materials and labor, may skew this total. On June 27, 2008, Dan Clery reported in the publication *Science* that "ITER scientists revealed a new cost estimate for the multibillion-dollar fusion reactor that was 30 percent higher than earlier calculations." Further budget adjustments will probably occur as construction gets started in the next few years. Project managers expect to achieve the first plasma experiment in 2025.

Despite its size and cost, ITER is not intended to be an economically viable reactor, but rather a stepping-stone toward this ambitious goal. Researchers and officials working on ITER believe that economic fusion power can be attained, if people are willing to invest in research that yields incremental advances. In an interview published in Clery's 2006 *Science* article, Lorne Horton at the Max Planck Institute for Plasma Physics in Germany said, "There's no doubt that it's an experiment. But it's absolutely necessary. We have to build something like ITER."

ITER's pending completion helps set goals for the continued JET experiments. Although much smaller, JET provides researchers with testing grounds for the effects of various currents and magnetic fields, with an eye toward performance improvements.

The ultimate goal is a power output that exceeds the input. But simply doing a little better than breaking even will not suffice for a viable economic power plant. Power plants such as nuclear (fission) reactors and coal- or oil-fueled utilities have many other costs, such as turning the energy into electricity, distributing the electric power, maintaining the facilities, and so forth. A moneymaking or at least a break-even fusion reactor must generate much more power than that required to heat and confine the plasma. The exact amount depends on engineering issues, but a successful reactor probably needs to amplify its power input by a factor of at least 30.

ALTERNATIVE APPROACHES TO FUSION—COLD FUSION

Fusion based on magnetic or inertial confinement can be accomplished in the laboratory, but no one knows if the techniques will ever lead to an economical means of generating power. The sticking point is the energy needed to produce the extremes in heat and pressure, which has led some researchers to wonder if those extremes are actually essential. Fusion in the Sun and other stars occurs in these conditions, but perhaps there are other reactions in which nuclei fuse under milder circumstances.

On March 23, 1989, two researchers announced in a highly publicized press conference in Salt Lake City, Utah, that they had achieved fusion reactions with a simple apparatus in the laboratory, conducted at room temperature. The University of Utah researcher Stanley Pons and his colleague Martin Fleischmann of Southampton University in the United Kingdom described an experiment in which electrodes—electrical *conductors*—immersed in heavy water created a process in which fusion occurred. Heavy water contains much deuterium, in the form of deuterium oxide (D_2O), as opposed to normal water, which mostly consists of the common hydrogen isotope ^1H and oxygen, H_2O. The study of electrodes and the reactions they initiate in various liquids or solutions is known as electrochemistry. When an electric current passed through the electrodes, which were made of palladium and platinum, Pons and Fleischmann claimed that the process generated more heat than would be expected from ordinary electrochemical reactions. That excess energy, the researchers said, came from fusion.

The announcement made headlines and shocked most theoretical physicists, who were convinced that fusion required high temperatures and pressures in order to overcome the electrical repulsion of nuclei. Pons and Fleischmann hypothesized that their simple experimental apparatus caused fusion to occur because the electric current split deuterium oxide and forced deuterium nuclei into tight spaces in the electrodes. Since high temperatures are not required, the process is called cold fusion—or, since the apparatus is small enough to fit on a laboratory bench (as opposed to the enormous facilities such as NIF and JET), scientists sometimes describe it as tabletop fusion. Fleischmann and Pons published their findings later in 1989.

Fusion events had been known to occur in experiments in which atoms are smashed together, which was how early physicists such as Oliphant discovered and studied this kind of reaction. These processes often take place in machines called particle accelerators that boost particles to enormous speeds, as discussed in chapter 2 of this volume. This subject is sometimes known as high-energy physics because the speeding atoms have a lot of energy, which can generate extraordinary re-actions, including fusion, when the atoms collide. But electrochemical fusion came as a surprise.

All scientific discoveries, and particularly surprising ones like cold fusion, must be confirmed with additional experiments, preferably ones conducted in a variety of different ways and by a variety of different ex-perimenters, some of which should be skeptical. A person who expects a certain result may not search diligently for alternatives, but skeptics will thoroughly explore other possibilities in order to explain something they find difficult to believe. When a skeptic replicates an experiment, the results tend to be more convincing.

In the case of cold fusion, confirmation did not occur. Because of the potential importance of this work, many laboratories attempted to replicate the results. A few succeeded to a certain extent, with various modifications, but most researchers failed. Most physicists came to be-lieve that the extra energy Pons and Fleischmann observed had nothing to do with fusion. Reinforcing this belief was the failure of physicists to find certain types of radiation or particles typically emitted during fusion events. In 1989, the Department of Energy (DOE) convened a panel of experts to review cold fusion research. The experts delivered their opinion in a 1989 DOE publication, *Cold Fusion Research,* stating that the results "do not present convincing evidence that useful sources of energy will result from the phenomena attributed to cold fusion. In addition, the Panel concludes that experiments reported to date do not present convincing evidence to associate the reported anomalous heat with a nuclear process."

But a few dogged researchers have continued this line of research. One of the most active is Peter Hagelstein at MIT. Hagelstein's persis-tence induced DOE to take another look at the topic, which they did in 2004. The opinion had not changed—the experimental results remained unconvincing.

In "Fusion as an Energy Source: Challenges and Opportunities," a report prepared for the Institute of Physics by W. J. Nuttall, cold fu-sion gets a cold shoulder, so to speak: "If cold fusion releases energy, as Hagelstein and others continue to report, then it does so without the production of large numbers of high-energy neutrons or other emit-ted reaction products. That would mean that the physics involved must differ fundamentally from that observed in a conventional 'hot fusion' process." The report goes on to summarize the present status of the field: "The orthodox view of cold fusion is that such phenomena do not exist. In response the proponents continue to suggest that such phenomena are merely difficult to generate."

"BUBBLE FUSION"

Most physicists continue to be skeptical about the prospects for tabletop fusion. A recent episode involving a process called "bubble fusion" highlights the reasons why.

In 2002, the researcher Rusi Taleyarkhan at Oak Ridge National Laboratory in Tennessee and his colleagues published a paper in *Science* in which they claimed to have observed fusion in a tabletop device. In this case, the device was a chamber filled with a liquid known as acetone, but with the hydrogen atoms replaced with deuterium. The experimenters created bubbles and then imploded them by sending high-pressure sound waves through the fluid. Using sound to manipulate bubbles or cavities in fluid is called acoustic cavitation. Although bubble implosions would not seem to be too violent, the abrupt movements are known in certain cases to create temperatures and pressures high enough to emit brief pulses of light, a process called sonoluminescence. Taleyarkhan and his coworkers claimed in their 2002 paper, "Evidence for Nuclear Emissions during Acoustic Cavitation," that the conditions were extreme enough that fusion was occurring, judging by the presence of typical deuterium fusion by-products such as neutrons and tritium.

Since the fusion reported in this experiment occurred during conditions of extreme pressures and temperatures—albeit of a highly transient nature and limited in spatial extent—the process was not the same as "cold" fusion. But the apparatus was small and fit in the tabletop category.

Similar to the experiments of Pons and Fleischmann 13 years earlier, bubble fusion got a lot of attention—and criticism. As the journalist Charles Seife wrote in the same 2002 issue of *Science,* "The heat from the controversy alone is nearly enough to trigger a nuclear reaction." In his article, Seife quoted Michael Saltmarsh, a physicist who is also employed at Oak Ridge National Laboratory and who disagreed that the emissions signaled the presence of fusion: "There's no evidence for any neutron excess due to fusion. If the tritium results in Taleyarkhan's paper are correct, and if you assume all the tritium is due to d-d [deuterium-deuterium] fusion, then you expect a 10-fold increase in the neutron signal. We see a 1 percent effect."

Taleyarkhan moved to Purdue University in Indiana two years later. He and his colleagues continued to study bubble fusion and produced results confirming and extending the original experiments. But other researchers could not. Their failures increased the belief that bubble fusion was not a real phenomenon and the emissions Taleyarkhan and his colleagues had seen were from other sources.

In 2006, frustrations grew to an intolerable level as researchers accused Taleyarkhan of hampering their efforts to evaluate his experiments. Purdue University decided to investigate the matter, and the journalist Robert F. Service reported in *Science* on March 17, 2006, that the university "announced that it was launching a review into allegations that Taleyarkhan—a nuclear engineer at Purdue and the field's chief proponent—had obstructed the work of Purdue colleagues by removing shared equipment, declining to share raw data, and trying to stop them from publishing results that countered his own published work."

This investigation, as well as one that followed, could not prove any serious wrongdoing. But members of a third panel said they found evidence of ethical lapses. As Service reported in *Science* on July 25, 2008, "After two previous investigations looked into alleged scientific misconduct by Taleyarkhan, a third panel has now cited Taleyarkhan for two cases of misconduct. Both cases centered on efforts by Taleyarkhan to make experiments carried out by members of his lab appear as independent verification of his previous work." In other words, the university charged Taleyarkhan with trying to make it seem that other researchers had confirmed his findings, which would have been more convincing than the efforts of his own coworkers. For instance, the investigatory panel concluded that Taleyarkhan had added and/or removed names on research papers so that the results would have appeared to come from another laboratory.

The investigation did not scientifically evaluate the bubble fusion experiments. But with the failures to replicate the findings, many researchers are skeptical. Once again, a claim for tabletop fusion seems, at least for the time being, to have fizzled.

CONCLUSION

Skepticism from the physics community would not prevent tabletop fusion researchers from building a reactor, if such techniques actually worked—a successful machine or operation does not require universal approval of scientists in order to function. But a lack of general approval usually limits the amount of funding a researcher working in an unpopular field might expect to receive. Although these limits could potentially hamper scientific advances, procuring the simple and relatively inexpensive equipment of tabletop fusion would not seem to offer any serious obstacles. If tabletop fusion techniques are valid, albeit difficult, then they will eventually offer great benefits. It is unfortunate that such an optimistic scenario seems a bit too good to be true.

NIF, JET, and the future ITER installation are expensive but necessary steps toward the possible development of fusion power, with all of its environmental advantages. No one is certain if inertial or magnetic confinement techniques will succeed in one day forming the basis of an economical fusion reactor, but the promise of a cheap, clean, and nearly boundless energy resource seems worth the risk.

The study of fusion also has benefits in and of itself. Nuclear processes are involved in radioactivity, particle physics, the evolution of stars, and other important branches of science. And much of the equipment used to study fusion can be put to a variety of uses.

One prominent example of a powerful piece of equipment is the Z machine at Sandia National Laboratories in Albuquerque, New Mexico. The Z machine can generate an enormous quantity of *X-rays*—high-frequency electromagnetic radiation that at small doses is useful in producing medical images, but has a lot of energy and is useful in studying fundamental processes in chemistry and physics. In a typical experiment, a huge amount of electricity, equal to several million amps of electric current, enters a target consisting of a metallic can containing a few hundred vertical tungsten wires. The wires are thinner than the diameter of a human hair and when hit with the powerful discharge they vaporize, forming a plasma. Magnetic fields arise from the currents and compress the plasma, similar to a Z-pinch. Since the axis of the target is vertical—the z-axis in terms of three-dimensional mathematical coordinate systems—the device is called the Z machine. The plasma implosion generates high temperatures as well as an extremely large amount of X-rays. During some experiments, the power output of these X-rays briefly exceeds the power of the world's supply of electricity by a factor of about 80.

In 2003, Sandia's Z machine managed to initiate fusion when researchers placed a small BB-sized deuterium pellet within the target region. This process is similar to that being studied at NIF. Experiments with the Z machine have also managed to melt a sheet of diamond, the hardest known natural substance, into a puddle.

Even greater temperatures have been achieved when scientists substituted larger wires made of steel for the tungsten arrays. These experiments generate temperatures exceeding 3,600,000,000°F (2,000,000,000°C). Such experiments could possibly succeed in fusing nuclei heavier than hydrogen isotopes.

Despite the progress, some scientists are pessimistic about "hot" fusion as much as they are about cold fusion, at least in terms of producing a viable reactor. The nuclear scientist William E. Parkins wrote an opinion column critical of fusion power research in *Science* on March 10, 2006. Parkins noted that, "In the early 1950s, the hydrogen bomb wakened public awareness to the explosive power of nuclear fusion and launched hope in the physics community to use fusion as a power source." Unlike fission, however, fusion has not been successful, for "although practical, controlled energy release from fission followed the discovery of that process by only 3 years, fusion power is still a dream-in-waiting." After chronicling the lack of success, Parkins gave a pessimistic summary: "The history of this dream is as expensive as it is discouraging."

Yet it is perhaps unrealistic to expect that an enormous technical and scientific advance such as fusion power should come effortlessly. The frontiers of physics are not for the faint of heart. No one can be sure of the outcome, but fusion experiments will lead to a better understanding of nuclear physics, and—possibly—to an almost limitless source of clean, cheap energy. Such an energy source would transform the world and alleviate a large fraction of the serious pollution and climate change issues that confront today's society. Considering the potential benefits, some researchers believe the effort is worth the risk of failure.

CHRONOLOGY

1860s   The Scottish physicist Sir William Thompson, Lord Kelvin (1824–1907), in accordance with earlier work by the German researcher Hermann Helmholtz (1821–94), hypothesizes that the Sun shines due to gravitational energy.

1896   The French physicist Henri Becquerel (1852–1908) discovers radioactivity.

1900s   The Polish scientist Marie Curie (1867–1934) and her husband, the French researcher Pierre Curie (1859–1906), study radioactivity and discover new radioactive isotopes.

1905   The German-American physicist Albert Einstein (1879–1955) discovers the formula $E = mc^2$, which establishes a relationship between energy and mass that is vital in nuclear processes such as fusion.

1911   The New Zealand-British physicist Ernest Rutherford (1871–1937) proposes the existence of the nucleus—a small, positively charged central region of the atom.

1932   The Australian physicist Sir Mark Oliphant (1901–2000) and his colleagues observe fusion reactions in hydrogen nuclei.

1938   The German-American physicist Hans Bethe (1906–2005) explains how nuclear fusion is the power source of the Sun and other stars.

The German scientists Otto Hahn (1879–1968) and Fritz Strassman (1902–80) discover a fission reaction in uranium.

1945   The United States uses two nuclear fission bombs—the first dropped on the Japanese city of Hiroshima on August 6 and the second three days later on the Japanese city of Nagasaki—to end World War II.

1950s Scientists in the United States and the former Soviet Union begin working on fusion power with the Z-pinch effect and magnetic confinement.

1952 The United States tests the first hydrogen bomb, a nuclear weapon employing the fusion of hydrogen nuclei.

1957 The first major nuclear reactor (based on fission) in the United States begins operation in Shippensport, Pennsylvania.

1979 The Joint European Torus (JET) project begins facility construction at Culham in the United Kingdom.

1983 JET construction is completed. The device achieves its first plasma.

1989 Stanley Pons and Martin Fleischmann announce that they achieved fusion with a simple electrochemical device operating at room temperature. The announcement creates a storm of controversy surrounding so-called cold fusion.

The Department of Energy convenes a panel of experts to investigate cold fusion claims. This panel concludes that cold fusion has not been convincingly demonstrated.

1997 Construction begins at the National Ignition Facility at Lawrence Livermore National Laboratory.

JET achieves 16 megawatts of fusion power.

2001 Construction of the NIF main building is completed.

2004 A second Department of Energy panel revisits the question of cold fusion but reaches the same negative conclusion as the first panel.

2006	The governments of China, India, Japan, South Korea, Russia, the European Union, and the United States formally agree to support ITER, an international project to build an experimental fusion reactor.
2009	NIF becomes fully operational.

FURTHER RESOURCES
Print and Internet

Bahcall, John N. "How the Sun Shines." Available online. URL: http:// nobelprize.org/nobel_prizes/physics/articles/fusion/. Accessed June 22, 2009. This splendid article, hosted at the Web site of the Nobel Foundation, explains the basics of fusion and how it works in the Sun and other stars.

Clery, Dan. "ITER Costs Give Partners Pause." *Science* 320 (6/27/08): 1,707. Rising costs are beginning to upset ITER participants.

———. "ITER's $12 Billion Gamble." *Science* 314 (10/13/06): 238–242. Clery describes ITER's hopes and challenges.

Department of Energy. "Cold Fusion Research." Available online. URL: http://www.ncas.org/erab/. Accessed June 22, 2009. The Energy Research Advisory Board to the DOE weighs in on the cold fusion controversy.

Einstein, Albert. "Does the Inertia of a Body Depend on Its Energy Content?" *Annalen der Physik* (Annals of Physics) 18 (1905): 639–641. Available online. URL: http://www.fourmilab.ch/etexts/einstein/E_ mc2/www/. Accessed June 22, 2009. This resource contains an English translation of Einstein's famous paper.

European Fusion Development Agreement. "Joint European Torus." Available online. URL: http://www.jet.efda.org/. Accessed June 22, 2009. This Web resource provides news and information on the latest research with JET, as well as articles on the basic principles of fusion.

Freudenrich, Craig. "How Nuclear Fusion Reactors Work." Available online. URL: http://science.howstuffworks.com/fusion-reactor.htm. Accessed June 22, 2009. The article, posted at the howstuffworks

Web site, describes the hypothetical operation of fusion reactors, assuming they will be developed along the lines of current research projects such as ITER.

Henderson, Harry. *Nuclear Physics.* New York: Facts On File, 1998. This book tells the fascinating story of the development of nuclear physics, focusing on the work of Marie and Pierre Curie, Ernest Rutherford, Niels Bohr, Lise Meitner, Richard Feynman, and Murray Gell-Mann.

Herman, Robin. *Fusion: The Search for Endless Energy.* Cambridge: Cambridge University Press, 1990. Although this book is dated—it was published shortly after Pons and Fleischmann announced their cold fusion experiments—Herman, a journalist, does a good job of describing the trials and tribulations of the quest for fusion power from the 1950s through 1990.

Intergovernmental Panel on Climate Change. *Climate Change 2007.* Available online. URL: http://www.ipcc.ch/ipccreports/ar4-wg2. htm. Accessed June 22, 2009. IPCC reports on the impacts of global climate change.

Isaacson, Walter. *Einstein.* New York: Simon & Schuster, 2007. Isaacson's biography includes papers that have only recently been released and paints a complete picture of the scientist who established much of the foundation of modern physics.

Kirkland, Kyle. *Atoms and Materials.* New York: Facts On File, 2007. Aimed at students in grades six–12, this book contains a discussion of atoms and molecules, including nuclear energy, and explains how states of matter and the properties of various materials depend on atomic physics.

Lawrence Livermore National Laboratory and the Princeton Plasma Physics Laboratory. "Fusion Energy Education." Available online. URL: http://fusedweb.pppl.gov/. Accessed June 22, 2009. This excellent educational Web resource features information on inertial confinement, magnetic confinement, the Sun, and other topics. A glossary, dictionary, and links to other Web sites are also offered.

Mackintosh, Ray, Jim Al-Khalili, Björn Jonson, and Teresa Peña. *Nucleus: A Trip into the Heart of Matter.* Baltimore, Md.: Johns Hopkins University Press, 2001. Beautifully illustrated and well written, this book examines the forces and structure of the atomic nucleus.

Nuttall, W. J. *Fusion as an Energy Source: Challenges and Opportunities.* September 2008. Available online. URL: http://www.ioppublishing.com/activity/policy/Publications/file_31695.pdf. Accessed June 22, 2009. This report offers an accessible and comprehensive discussion of the efforts to build a fusion reactor.

Parkins, William E. "Fusion Power: Will It Ever Come?" *Science* 311 (3/10/06): 1,380. In this skeptical editorial, Parkins notes the steep challenges faced by fusion researchers.

Rygg, J. R., F. H. Séguin, C. K. Li, J. A. Frenje, et al. "Proton Radiography of Inertial Fusion Implosions." *Science* 319 (2/29/08): 1,223–1,225. The researchers adapt radiography in order to take pictures of the activity within small fuel pellets as they implode.

Seife, Charles. "'Bubble Fusion' Paper Generates a Tempest in a Beaker." *Science* 295 (3/8/02): 1,808–1,809. This news article accompanies the Taleyarkhan paper in the same issue.

———. *Sun in a Bottle: The Strange History of Fusion and the Science of Wishful Thinking.* New York: Viking, 2008. As the title suggests, the science journalist Charles Seife casts a skeptical eye on the possibility of developing economical techniques of harnessing fusion power, both the hot and cold variety.

Service, Robert F. "New Purdue Panel Faults Bubble Fusion Pioneer." *Science* 321 (7/25/08): 473. An investigatory committee at Purdue University cites the bubble fusion researcher Rusi Taleyarkhan for misconduct.

———. "Researchers Raise New Doubts About 'Bubble Fusion' Reports." *Science* 311 (3/17/06): 1,532–1,533. Some physicists question bubble fusion research.

Taleyarkhan, R. P., C. D. West, J. S. Cho, R. T. Lahey, Jr., et al. "Evidence for Nuclear Emissions during Acoustic Cavitation." *Science* 295 (3/8/02): 1,868–1,873. The researchers claim to have observed fusion in a tabletop device.

Taubes, Gary. *Bad Science: The Short Life and Weird Times of Cold Fusion.* New York: Random House, 1993. Taubes, a journalist, documents the tremendous excitement that Pons and Fleischmann's 1989 announcement made and the acrimonious controversies that followed.

Walter, Katie. "The National Ignition Facility Comes to Life." *Science and Technology Review* (September 2003). Available online. URL: https://www.llnl.gov/str/September03/Moses.html. Accessed June 22, 2009. This article offers an excellent introduction to NIF.

Wijers, Ralph. "Obituary: Hans Albrecht Bethe, 1906–2005." *Bulletin of the American Astronomical Society* 39 (2007): 1,055. Wijers marks the passing of this highly respected physicist.

Web Sites

ITER. Available online. URL: http://www.iter.org/. Accessed June 22, 2009. ITER's site offers the latest news on the development and progress of this ambitious fusion project.

National Ignition Facility & Photon Science. Available online. URL: http://lasers.llnl.gov/. Accessed June 22, 2009. NIF's Web site provides news and information on the world's largest laser system. A photo gallery, video gallery, and virtual tour are included.

Princeton Plasma Physics Laboratory. Available online. URL: http://www.pppl.gov/. Accessed June 22, 2009. This laboratory is one of the most important fusion research centers in the world. The Web site describes fusion processes, the potential of fusion power, and the current projects at the laboratory.

2

PARTICLE ACCELERATORS

Researchers study tiny objects by "illuminating" them with certain forms of energy, such as light if the object is big enough to be seen with visible light, or if not, with sources such as X-rays. X-rays have a much smaller *wavelength* than visible light and can discern finer details, similar to the way a thin probe reveals small nooks and crannies that would be missed by a probe that is too large to fit. X-rays can penetrate certain materials, such as the soft tissue of the human body, to reveal the structure and health of bones, and scientists use a technique known as X-ray crystallography to determine the structure of proteins and other important biological molecules.

In the study of atomic particles, physicists need an even more powerful technique. This is the role of particle accelerators. The enormous energy of these machines gives scientists an extraordinary view of atomic and subatomic phenomena.

To achieve these enormous energies, particle accelerators must be big—and expensive. Experiments with particle accelerators have given physicists insight into the smallest components of matter and the laws of physics at these tiny scales, but the price tag has been stiff. The most powerful particle accelerator in the world, the Large Hadron Collider (LHC), cost about $6 billion. This chapter discusses these giant machines and how they work, what physicists who operate them have discovered, and what more there is to learn at this subatomic frontier of physics. Particle accelerators offer one of the few avenues by which physicists can study the basic laws of physics underlying the universe, as well as the intriguing mystery of how these laws, and the universe itself, came to be.

INTRODUCTION

Devices called doomsday machines are sometimes featured in exciting though far-fetched stories depicting end-of-the-world scenarios. Such stories have proliferated since the development of nuclear weapons in the 1940s, which is understandable considering the destructive potential of these bombs. Although even the largest nuclear weapon presently available is not powerful enough to destroy the world, the collective arsenals of nuclear-armed countries could do so. These stories involve a single powerful device capable of ending civilizations or demolishing planets, as in "The Doomsday Machine" episode in the original *Star Trek* series, the 1970 movie *Beneath the Planet of the Apes,* and the satirical 1964 movie *Dr. Strangelove or: How I Learned to Stop Worrying and Love the Bomb.*

Perhaps the popularity of these fictional accounts explains some of the negative publicity and criticisms of the Large Hadron Collider. LHC is a particle accelerator, boosting particles called *hadrons* to enormous speeds and studying the aftermath of collisions. Situated at the border between France and Switzerland, LHC was completed in 2008 and is the world's largest particle accelerator. But prior to LHC's initial operation on September 10, 2008, some people were worried. Fears included the creation of a *black hole*—an object with a gravitational field so powerful that nothing, not even light, can escape it—and other exotic phenomena. As the reporters Dan Clery and Adrian Cho wrote in *Science* on September 5, 2008, "A handful of physicists and others have proposed an array of dangerous entities that could be created in the minuscule fireball of a particle collision—including microscopic black holes, strange matter that is more stable than normal matter, magnetic monopoles, a different quantum-mechanical vacuum, and even thermonuclear fusion triggered by a stray beam. Discussion forums on the World Wide Web sizzle with rants against arrogant scientists who meddle with nature and put us all at risk." A few people even filed lawsuits, which the courts dismissed.

This adversity occurred despite attempts of scientists associated with LHC to quell it. The European Organization for Nuclear Research

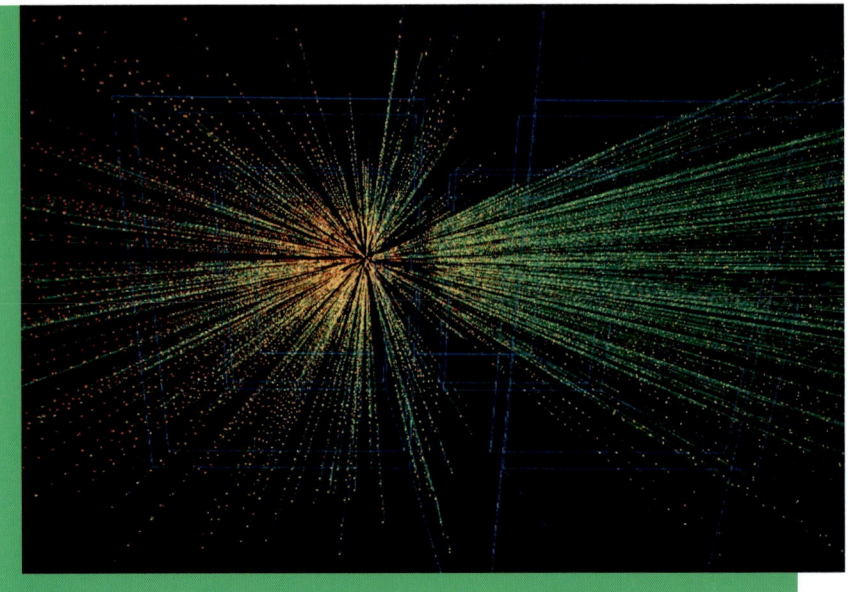

Particle tracks resulting from lead ion collisions at CERN *(CERN)*

(CERN) operates LHC. (The acronym—CERN—is based on an old French name, *Conseil Européen pour la Recherche Nucléaire* [European Council for Nuclear Research]. CERN kept its well-known acronym even after changing its name.) CERN scientists conducted a study in 2003 that showed fast-moving particles called *cosmic rays* collide with particles in Earth's atmosphere with even greater energy than they would in LHC. If such collisions were powerful enough to bring about doomsday, they would have already done so.

Worries about the risks associated with LHC operation illustrate the confusion and debate over the usefulness of these high-energy devices. To study the fundamental properties of matter and the universe, particle accelerators—sometimes known as atom smashers—must generate extraordinary energies, similar to the fusion devices discussed in chapter 1. Questions over whether such projects are worthwhile have tightened the budgets of particle accelerators; in 1993, for example, the U.S. government canceled a large-scale project known as the Superconducting Supercollider after more than $2 billion had already been spent.

Extreme energies are essential to modern physics for several reasons. One reason involves the principles of *quantum mechanics,* which govern the behavior of atomic and subatomic particles. Although it sounds strange, all objects have wavelike and particle-like properties according to quantum mechanics. For example, light can appear as a particle called a photon or as an electromagnetic wave, depending on how scientists measure it. The wavelength of an object depends on energy—smaller wavelengths correspond to higher frequencies and higher energies. These shorter wavelengths can probe the small structures of atomic and subatomic particles.

Another reason for high energies is revealed in the equation $E = mc^2$, concerning energy, (E), mass (m), and the speed of light in a vacuum (c). As described in the sidebar "Einstein's Famous Equation: $E = mc^2$" on page 6, Albert Einstein used this equation to show that energy and mass are related. This equation is important in nuclear reactions because a portion of the mass is transformed into energy. In particle physics, the equation is also important for the opposite transaction. Many particles are short-lived and must be continually created (after which they promptly decay into other particles), and creating these particles requires energy. In particle accelerators, some of the energy of the colliding particles is transformed into mass in the form of rarely seen particles, allowing physicists a chance to study them. According to Einstein's equation, a particle of mass m is equal to E/c^2. Because c is so large—about 186,000 miles per second (300,000 km/sec)—E must be large as well. High energies are essential in these experiments.

Early experiments in atomic physics involved particle accelerators, although the equipment was not referred to as such at the time. In 1897, the British physicist Sir Joseph John Thomson (1856–1940) studied a beam of particles, accelerated with electricity, in which he discovered the electron.

The New Zealand-British physicist Ernest Rutherford (1871–1937) and his colleagues discovered the atomic nucleus in 1911, during experiments in which they used radioactive emissions to bombard extremely thin sheets of gold. In this experiment, radioactive isotopes emitted energetic particles known as alpha particles, consisting of two protons and two neutrons (or, in other words, the nucleus of the common helium isotope ^4He). Alpha particles have a positive charge, and Rutherford expected the thin sheet of gold, which was only a few atoms thick, to deflect some of them as they encountered charges such as the electron, which Thomson had discovered a few years earlier. But Rutherford was shocked to find that in some cases the alpha particle deflected backward, as if it had run into the wall—or a dense, positively charged particle that repelled it. From these experiments, Rutherford deduced the existence of a congregation of positively charged particles in the gold atoms—the nucleus, which contains positively charged protons and neutrons.

In addition to radioactive emissions, early particle physicists made use of cosmic rays. These rays consist mostly of high-speed protons, the source of which is not yet fully understood. Many of these protons reach Earth and can provide an experimenter with a lot of high-speed particles. But the planet's atmosphere scatters most of these particles, and few reach the surface, so desperate experimenters of the early 20th century climbed a mountain or ascended in a balloon, along with their equipment, to take advantage of this resource.

The desire to control the particle beam and to create high-speed particles at will, rather than relying on dangerous, sporadic radioactive emissions or difficult-to-reach cosmic rays, motivated the development of modern particle accelerators. Perhaps the simplest particle accelerator is the cathode ray tube (CRT), in which an electric potential accelerates charged particles emitted from one plate, the cathode, to another plate (at higher potential), called the anode. This device was the basis of early television sets and computer monitors and is still often used for this purpose, although newer technologies such as liquid-crystal displays and plasma TVs are replacing it. A CRT was also the basis of Thomson's electron experiments, before which the emissions were referred to as rays rather than beams of particles, which accounts for the *ray* term in the name.

To accelerate particles to greater speeds and thus greater energy, researchers needed increasingly high voltages. In 1929, the American physicist Robert Van de Graaff (1901–67) designed a generator that transfers a huge amount of charge to a metal globe, which develops a large potential difference. This potential difference, measured in volts, applies a force to charges, as in CRTs. A common flashlight battery uses chemical reactions to generate about 1.5 volts, but Van de Graaff generators can produce thousands and even millions of volts. The British physicist Sir John Cockcroft (1897–1967) and the Irish physicist Ernest Walton (1903–95) designed a different high-voltage system consisting of components that multiplied voltage. In 1932, Cockcroft and Walton used this system to build an accelerator at Cambridge University in the United Kingdom.

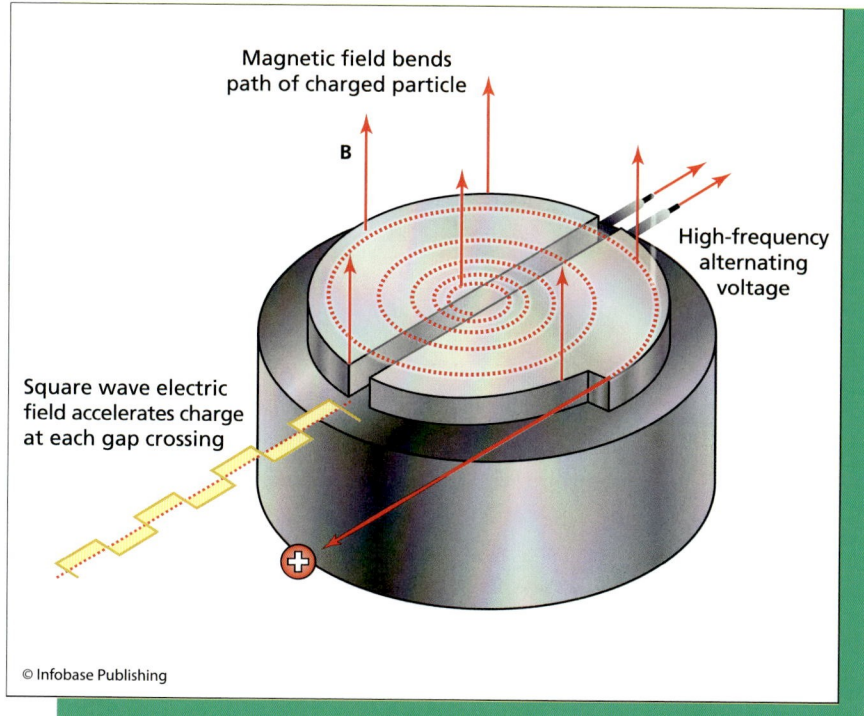

Magnetic field bends
path of charged particle

B

High-frequency
alternating
voltage

Square wave electric
field accelerates charge
at each gap crossing

© Infobase Publishing

In the simple cyclotron illustrated here, hollow D-shaped plates hold alternating voltages. A magnetic field (*B*) curves the motion of charges traveling inside, as they pick up speed and shoot out of the exit site.

But using increasingly high voltages is dangerous. In a linear accelerator such as the Cockcroft-Walton machine, the high voltage between two points accelerates the particles in a straight line, as they travel from one plate to another. The voltages can be divided into segments, but if the number of segments is large, the length of the accelerator will be excessive (and extend well beyond the laboratory!). In the late 1920s, the American physicist Ernest O. Lawrence (1901–58) at the University of California, Berkeley, began developing the means of applying a smaller accelerating force—a smaller voltage, in other words—but doing so repeatedly by sending the particles through a loop. By 1930, Lawrence had built a machine called a *cyclotron* that used strong magnetic fields to apply a force to the particles, which move in a circle because the magnetic force acts perpendicular to the direction of motion. (See the section "Magnetic Confinement—A Bottle with No Walls" in chapter 1.) The particles pick up speed as they spiral outward, eventually shooting out at high velocities. Lawrence referred to the machine as a "proton merry-go-round." Although modern accelerators are more complicated, they continue to use magnetic fields and high voltages to steer and accelerate particles.

ACCELERATING A PARTICLE TO NEARLY THE SPEED OF LIGHT

Despite the high voltages of linear accelerators and the looping action of circular accelerators, there is a speed limit beyond which particles cannot go. This speed limit is c, the speed of light in a vacuum, which is 186,000 miles per second (300,000 km/sec). No particle can be accelerated beyond this velocity.

Why is c the speed limit? One way of looking at it is to consider Einstein's special theory of relativity, which holds that particles with

A synchrocyclotron at the Joint Institute for Nuclear Research (JINR) in Dubna, Russia *(JINR)*

mass seem to increase their mass with increasing velocity. The formula Einstein derived for the relativistic mass, m_r, and the object's velocity, v, and resting mass, m, is $m_r = m/\sqrt{(1 - v^2/c^2)}$. This means that the relativistic mass equals the resting mass, m, divided by the square root of 1 minus the square of v divided by c. When $v = 0$, the formula reduces to $m_r = m$, which makes sense—when the velocity is zero, relativistic mass equals resting mass. For low velocities compared to the speed of light—and the fastest jet airplanes move slowly compared to light—m_r is not much different than resting mass, as the reader can demonstrate by plugging in a few numbers. But as v comes close to c, the mass rises rapidly. When $v = c$, the denominator of the left side of the formula goes to zero, which is a mathematical impossibility (division by zero makes no sense). Einstein concluded that no object with mass can be accelerated up to c, and subsequent observations in particle accelerators have borne out this conclusion.

This increase in mass causes a problem for the cyclotron, since the properties of the magnetic field depend on mass. A machine known as a synchrocyclotron takes the relativistic corrections into account, compensating for the increases in mass.

Even though particles cannot be accelerated up to c—or beyond, of course—they can reach speeds arbitrarily close. The particles travel so fast that their path must be a vacuum, since collisions with air molecules would scatter the beam. In most modern particle accelerators, particle speed can reach more than 99.99 percent of the speed of light or even greater, but physicists tend to think in terms of energy instead of velocity, which makes sense because of Einstein's equations and the relation between energy and mass. The standard unit of energy in particle physics is the *electron volt,* abbreviated eV, and equal to the kinetic energy gained by an electron as it accelerates through a potential difference of one volt. One eV is tiny in everyday applications; for instance, a 60-watt lightbulb uses 3.75×10^{20} eV in one second—375 million trillion eV! But to tiny particles such as electrons, eV is an appropriate unit. In terms of an electron or proton, a billion eV is an enormous quantity energy.

Two of the most important particle accelerator laboratories in the United States are the SLAC National Accelerator Laboratory and the Fermi National Accelerator Laboratory (commonly known as Fermilab). SLAC, operated by Stanford University, was formerly known as the Stanford Linear Accelerator Center, which accounts for the acronym. There

An aerial view of SLAC's layout *(SLAC)*

are multiple accelerators at SLAC, including a linear accelerator whose length is 2 miles (3.2 km), the longest linear accelerator in the world. Fermilab, located in Batavia, Illinois, is home to the second largest particle accelerator, the Tevatron, a circular accelerator. The sidebar on page 43 provides more information on Fermilab.

In order to study these particles, physicists not only need to accelerate them to high energies, they also need to observe the results. Atomic and subatomic particles are far too small to be viewed by conventional microscopes, and instruments such as the scanning tunneling microscope, which is capable of "seeing" atoms, will not work with high-speed particles and their collisions. Researchers usually visualize the particles from the effects that occur as the particles bump into or fly past molecules of the detector. For instance, Geiger counters are simple particle detectors that sense an increase in electrical conductivity that occurs when high-speed particles cause ionization—the particles rip or knock electrons from the detector's atoms and molecules, resulting in ions that conduct electric current.

Fermi National Accelerator Laboratory

The Italian physicist Enrico Fermi (1901–54) played an instrumental role in the development of nuclear weapons as well as nuclear reactors. After winning the Nobel Prize in physics in 1938 for his studies of neutrons and nuclear processes, Fermi fled Italy and its oppressive dictator, Benito Mussolini, and came to the United States. At the University of Chicago, in courts where the game of squash was once played under the university's Stagg Field, Fermi and his colleagues built the first nuclear reactor in 1942. In 1944 Fermi became an American citizen and accepted a permanent job at the University of Chicago, where he continued his studies of nuclear reactions and high energy physics. Fermi National Accelerator Laboratory, or Fermilab, was renamed in 1974 to honor the contributions of this physicist.

Fermilab got its start as the National Accelerator Laboratory in 1967. The goal was to develop a huge circular accelerator, which began operation in 1972. Building upon this early facility, the Tevatron was completed in 1983. Tevatron's ring is four miles (6.4 km) in circumference, which until the completion of LHC was the largest accelerator in the world. (The circumference of LHC is 16.6 miles [26.7 km].) The Tevatron can accelerate particles up to one trillion eV; the abbreviation for a trillion eV is TeV, which forms the basis of the initial portion of the Tevatron's name. A consortium of 90 research universities runs the laboratory.

Such huge equipment requires a substantial amount of space. Fermilab is situated on about 10 square miles (256 sq km) of ground, and it is used wisely. Originally farmland and woods, laboratory personnel have conserved many of the natural features of the area and, in 1969, introduced a small herd of North American bison onto the property.

(continues)

(continued)

Fermilab researchers have made many important contributions to particle physics, including the discovery of several *quarks,* which are fundamental particles that are the constituents of protons and neutrons. But CERN's Large Hadron Collider, completed in 2008, will assume many of the duties Tevatron once performed. The loss of this research, and the funding that supports it, has resulted in serious budget cuts at Fermilab. Officials are working to attract new projects to keep this important national laboratory at the frontiers of physics.

More sophisticated instruments include cloud chambers, in which particles leave tracks of visible "clouds" of water droplets that form around ionized atoms, and bubble chambers, in which the particle tracks are due to bubbles of vaporized fluid. But the detectors used most often in today's instruments are elaborate circuits of highly sensitive electronic instruments. Because some of the particles are not easy to find, particle detectors can be huge. One of the LHC detectors, known as Atlas, is 151 feet (46 m) long, 82 feet (25 m) high, and 82 feet (25 m) wide—the volume of a mansion!

The magnetic fields required for these powerful accelerators and detectors are enormous. Iron magnets are not usually sufficient or would be too large, so researchers use superconducting magnets. A superconductor is a conductor that has no electrical *resistance* and is the subject of a great deal of research, as described in chapter 4 of this book. Electric currents generate magnetic fields, but resistance reduces the current and generates heat. The absence of resistance in superconductors means they can generate magnetic fields at higher intensities and more cheaply than other magnets, although superconductors require elaborate cooling systems. Tevatron, LHC, and other large particle accelerators would not be possible without superconductors.

These pipes carry beams of particles in an accelerator at SLAC.
(Stanford Linear Accelerator Center)

Physicists use particle accelerators and the associated detectors to probe the fundamental components of matter, create rare or exotic particles, and study particle forces and interactions. But accelerators also have other applications, especially in the field of medicine.

ACCELERATORS IN MEDICINE

Cancer is a disease in which certain cells experience runaway growth and invade other tissues and is the second leading cause of death in the United States after heart disease. Treatments involve killing the cancerous cells while leaving healthy cells uninjured, although sparing healthy cells is sometimes difficult because cancerous cells often spread throughout the body. To kill the cancerous cells, physicians use chemicals in some cases, but in other cases they turn to various forms of ionizing energy. Ionization treatment is effective because it damages important biological molecules such as deoxyribonucleic acid (DNA), which kills cells, and cancerous cells tend to be more susceptible than healthy ones.

Patient undergoing neutron therapy at Fermilab *(Fermilab Visual Media Services)*

Some treatments use radioactive materials or high-energy electromagnetic radiation such as X-rays, but other treatments involve high-speed particles. Valuable sources of these particles are accelerators.

For example, some hospitals are now using beams of protons to irradiate and kill cancerous cells. Protons are generally more controllable than other ionizing radiation, such as X-rays, since the particles' speed and how far they travel in the body can be adjusted with the accelerator controls. Physicians can adapt the proton beams to suit specific cases, which may help them design effective treatments while reducing unwanted side effects, such as the loss of healthy cells. But proton therapy is expensive, since the treatment facilities must have access to a particle accelerator (though it need not be nearly as large as the machines at major research centers). Common accelerators for this purpose are small cyclotrons, synchrocyclotrons, or a circular accelerator known as a synchrotron, which is similar to a cyclotron but is also often used to generate X-rays as well.

Certain types of cancers resist standard treatment attempts, in which case physicians must seek alternatives. One alternative is a neutron beam. Instead of killing cancerous cells with ionization, neutrons tend to attack these cells by engaging in nuclear reactions (which also makes neutrons vital in nuclear reactors). Fermilab houses the Northern Illinois Institute for Neutron Therapy, one of the few neutron therapy centers in the United States.

But how is a neutral particle such a neutron accelerated? The methods described above use various electromagnetic properties to apply forces to charged particles, but they will obviously not work for a neutron. Yet generating a neutron beam is possible, though trickier. One technique temporarily attaches the neutrons to protons, speeds up the combination with a standard accelerator, and then detaches the neutrons "on the fly." Another technique generates high-speed neutrons by bombarding a target with other particles, producing collisions that send neutrons skittering away; narrow tubes or openings shape the neutron "debris" into a beam. Charged particles in the debris can be filtered out with the application of magnetic fields.

THE NATURE OF MATTER

Practical applications such as cancer therapies are important, but physicists are also interested in studying the laws of physics and the nature of matter. Particle accelerators let researchers probe into the very heart of physics.

Many different particles emerged from experiments beginning in the 1930s, and the number increased as particle accelerators became more powerful. Important properties of a particle include its charge (magnitude and sign), mass, and a quantum mechanical aspect known as *spin* (which is complex and is not quite equivalent to rotation, despite its name). These particles interact with one another in various ways—attracting, repelling, or combining—and can be classified by the kinds of forces their interactions employ. The four forces are gravitational, electromagnetic, *weak nuclear force* (or *weak force*) that governs certain nuclear interactions, and the strong nuclear force (or strong force) that governs other nuclear reactions, such as the proton-proton attraction in the nucleus, as described in chapter 1. Particles known as *leptons* engage in electromagnetic and weak interactions, while hadrons interact with the strong force. Gravitational forces are so weak on the particle level that they have played little role in particle physics.

Leptons include electrons and a slightly heavier particle called a muon, plus the neutrinos, which are mysterious particles emitted in many nuclear reactions. Lepton comes from the Greek word *leptos,* meaning "small," which aptly describes these lightweight particles. Hadrons include protons, neutrons, a class of particles called mesons, and other particles that tend to be much heavier than leptons—hadron comes from the Greek word *hadros,* which means "thick."

As dozens of different types of particles emerged from accelerator experiments, physicists struggled to make sense of this particle "zoo." A similar situation occurred much earlier in the history of science, when chemists tried to understand the varied properties of the chemical elements and finally succeeded when the Russian chemist Dmitry Mendeleyev (1834–1907) constructed the periodic table of chemical elements in the 1860s.

Great progress in particle physics emerged in 1963 when the California Institute of Technology researchers Murray Gell-Mann and George Zweig independently postulated the concept of quarks. (Gell-Mann adapted the term *quarks* from James Joyce's *Finnegans Wake,* which contains the line, "Three quarks for Muster Mark.") Quarks are fundamental particles, meaning that—at least according to physicists' present understanding—they are not composed of other, smaller particles. Quarks come in six "flavors" or types—up, down, strange, charm, bottom, and top—and combine to form hadrons. For example, a proton consists of two up quarks and a down quark. Leptons are not composed of quarks and appear to be fundamental particles as well.

Proving the existence of quarks was not easy. The strong force binds these particles so tightly that no free quark has ever been observed. But in the late 1960s and early 1970s, researchers at SLAC performed experiments similar to those that Rutherford had used to find the atomic nucleus. A team of researchers led by Richard Taylor at SLAC and Jerome Friedman and Henry Kendall at the Massachusetts Institute of Technology (MIT) sent a beam of electrons hurtling into protons and neutrons in the 2-mile (3.2-km) linear accelerator. A surprisingly large fraction of these electrons bounced off at extreme angles, suggesting that the protons and neutrons were not a uniform sphere, but instead consisted of pointlike-particles—quarks.

Finding the heaviest quark, top, was the job of Fermilab's Tevatron. The accelerator was just able to generate enough energy to create conditions that, every once in a while—once in a few billion collisions—showed the "signature" of a top quark. This signature is not an isolated particle, but instead is a set of other particles—the top quark decays at once into other particles that make certain tracks in the detectors. But since these tracks can be confused with others unrelated to top quark decay products, physicists had to pore over a huge quantity of data, involving billions of experiments, to find enough evidence to confirm the existence of the top quark. (Researchers can initiate particle collisions but have little control over what particles are created.) In 1995, two teams of researchers, each team composed of about 450 members, announced that they had found the top quark. The achievement took years and the efforts of a lot of people.

Particle detectors were responsible for another critical discovery in physics—*antimatter*. In 1932, the American physicist Carl Anderson (1905–91) studied some tracks in a cloud chamber made during collisions involving cosmic rays. He found a particle with the same mass as the electron but with a positive charge. This particle, called a positron, is the electron's *antiparticle*—a particle having the same mass and spin but opposite electric charge (and a few other properties). The British physicist Paul Dirac (1902–84) had postulated the existence of this particle in 1928 while working on some mathematical formulas of quantum mechanics.

Particle physicists eventually discovered that every particle has an antiparticle. (A few particles, such as the photon, are their own antiparticle.) The antiparticle of a proton, for example, is known as an antiproton. These antiparticles are examples of antimatter, which is similar to matter except for opposite properties. As described in the sidebar on page 50, matter and antimatter are annihilated when they come into contact, and the combined mass is turned into energy by Einstein's formula, $E = mc^2$.

The Energy of Matter-Antimatter Reactions

When matter and antimatter come into contact, they disappear in a burst of energy, often in the form of photons, the particles of electromagnetic radiation. The amount of mass dictates the energy of the photons. Electron-positron pair annihilation, for example, creates two photons with a total energy of 1.022 million eV (or 1.022 MeV, where M stands for mega). The heavier proton-antiproton pair annihilation yields much more energy: 1.88 billion eV (or 1.88 GeV, where G stands for giga). The inverse process also occurs, which is called pair production. In pair production, energy is transformed into a particle-antiparticle pair. For the electron-positron pair, at least 1.022 MeV is needed. (Any extra energy in the process is transformed into kinetic energy, or in other words, motion of the resulting particles.) Proton-antiproton pair production requires at least 1.88 GeV. Only the most powerful accelerators, such as Tevatron and LHC,

Particle accelerators are essential to study antimatter because they provide the energetic collisions necessary to create antiparticles. Antimatter has not yet been found outside of the laboratory, despite extensive searches. Although this is perhaps not too surprising—the universe contains much matter, which is annihilated, along with antimatter, when the two meet—it raises the question of why there is such an imbalance, or asymmetry. If the universe contained equal amounts of matter and antimatter in its early development, both would have been wholly consumed, leaving nothing but energy. Only by postulating a slight excess of matter at the very beginning can physicists account for the presence today of plenty of matter and no antimatter, except for a few antiparticles created during accelerator experiments.

can produce enough energy to create the larger particle-antiparticle pairs.

Antimatter is similar to matter except for the opposite properties, and anti-atoms made of antiprotons and positrons are possible. In 1995, Walter Oelert and his colleagues at CERN fired antiprotons at xenon atoms that resulted, in rare cases, in an atom of anti-hydrogen—a positron and antiproton—the first such observation. The anti-hydrogen atoms lived for about 40 billionths of a second before getting annihilated.

Since all or most of the mass is converted into energy, matter-antimatter annihilation would be an extremely efficient method of creating energy, such as needed to power a spaceship (which in science fiction sometimes has an "antimatter" engine). One problem with this idea is keeping the antimatter away from the matter until the energy is needed. Another problem is obtaining the antimatter. Particle accelerator experiments that produce antiparticles are not cheap; with current technology, the process of obtaining antiprotons costs about \$62.5 trillion per 0.035 ounce (1 g). This probably makes it the most expensive substance on Earth!

Particle physicists are actively studying this problem. An ongoing series of experiments at SLAC called BaBar, involving about 600 physicists, is attempting to find differences, or asymmetries, between particles known as B mesons and their antiparticle. (The name of the project is based on the representations B and B-bar for B mesons and anti-B mesons, respectively, as well as a sly reference to the fictional elephant Babar.) The Iowa State University physicist Soeren Prell noted in a November 13, 2008, press release issued by the university, "We found that B mesons and anti-B mesons behave differently." But the differences are complex, and further analysis is needed in order to understand how these differences may have led to the dominance of matter early in the universe's evolution.

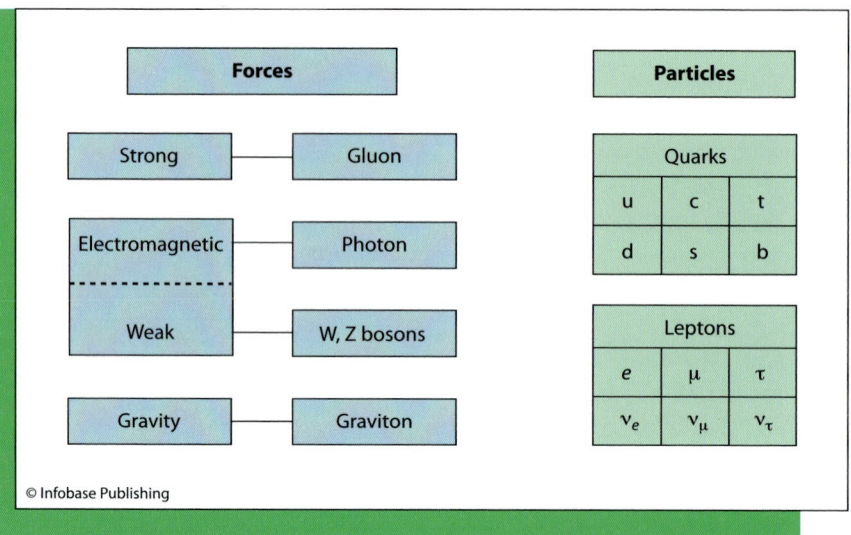

The standard model consists of four forces and their carrier particles, listed on the left, and 12 fundamental particles, listed in symbols on the right. Quark types are up, charm, top, down, strange, and bottom; the leptons are electrons, muons, taus, electron neutrinos, muon neutrinos, and tau neutrinos.

STANDARD MODEL OF PARTICLES AND INTERACTIONS

Accelerator experiments on quarks, antiparticles, leptons, and how they interact have given physicists insight into the fundamental laws of physics. The result is a theory known as the standard model that explains the nature of matter and energy in terms of 12 particles and three forces—electromagnetic, weak, and strong—and can be hypothetically extended to include the force of gravitation.

The 12 particles are six quarks, as given above, and six leptons, which include the electron, muon, tau, and three varieties of neutrinos (which are the subject of chapter 3). Governing the behavior of these particles are the four forces. According to the theory, certain particles mediate, or carry, these forces. The familiar photon carries the electromagnetic force, while a particle known as a gluon carries the strong force, and W and Z *bosons* carry the weak force. (Bosons are particles that have certain spin values and behave according to the advanced formulations of the Indian physicist Satyendra Nath Bose [1894–1974] and Albert Einstein.) A hypothetical particle called graviton mediates the gravitational force, assuming the standard model applies to all of the four known forces. (The force of gravitation is generally negligible between small particles—the other forces are much stronger—so particle physicists have not been able to explore this force in detail.) The figure on page 52 illustrates the elements of the standard model.

What role do the carrier particles play in effecting their respective forces? The answer belongs in the advanced realm of quantum mechanics, which is compatible with but on a different conceptual level than classical concepts such as fields. Physicists often speak of electric, magnetic, and gravitational fields to explain how large objects exert forces on one another, and these fields are a convenient description of the interactions. But particle physics theory indicates that particles experiencing a force will exchange the appropriate carrier particles, whose existence is fleeting or "virtual" but allowed by unfamiliar but demonstrable concepts of quantum mechanics.

The standard model is an important achievement and reduces the huge number of particles and interactions to the behavior of only a few fundamental components. An important set of predictions of the model, confirmed by experiment, is the existence and masses of the W and Z bosons. In 1983, the CERN physicist Carlo Rubbia and his colleagues announced they had observed these particles in their high-energy experiments. These particles are quite massive—the W boson has the mass of about 80 GeV/c^2 (80 billion eV divided by the square of the speed of light) and Z about 91 GeV/c^2. For the sake of comparison, the proton is about 0.94 GeV/c^2.

One of the puzzling features of the group of carrier particles is why the W and Z bosons are so massive, while the photon has zero mass. This difference underlies the different properties of the weak force (mediated by W and Z) and the electromagnetic force (mediated by photons). Yet in high-energy conditions, such as during violent collisions in particle accelerators, these two forces share the same formulas, and the combination is called the electroweak force. The disparity between electromagnetism and the weak force at low energies has led researchers to postulate the existence of a boson called the Higgs boson, named after the British theoretical physicist Peter W. Higgs, who proposed it.

The Higgs boson works by a complicated mechanism that exerts a kind of field that resists particle motion and provides them with inertia—a manifestation of mass. As a result of this mechanism, the Higgs boson accounts for the differences in the electromagnetic and weak forces and the masses of their carriers. This particle also explains the masses of other particles. The Higgs boson is so instrumental in the standard model that physicists sometimes refer to it as the God particle.

One big problem with the Higgs boson is that physicists have not yet observed it—and therefore have not yet proved that it exists—despite elaborate searches. A major obstacle of these searches is that although the Higgs boson explains the mass of other particles, theorists do not know what the mass of the Higgs boson should be. Since it has yet to be found, physicists assume the mass is large and therefore requires a tremendous amount of energy to produce, perhaps beyond the range of earlier machines. CERN physicists conducted a series of experiments and concluded in 2000 that the Higgs boson could not have a mass of less than about 115 GeV/c^2.

Researchers are now closing in. Fermilab scientists announced in 2008 that they had raised the lower limit of the Higgs boson's mass to about 170 GeV/c^2. The researchers achieved this result by using Tevatron to produce pairs of Z bosons. This pair production requires enormous energy, and the characteristics of the experiment are similar to that which would observe the hypothetical Higgs boson. Even with the Tevatron, Z boson pair production is an exceedingly rare event. To observe these events, a team of about 600 physicists from 18 countries inspected the outcome of almost 200 trillion particle collisions.

Because the Higgs boson's mass is so large, researchers are now pinning their hopes on the LHC. In addition to the ability to generate the highest energies yet attainable, LHC also has sensitive detectors such as the Atlas that key in on the possible decay products of the Higgs boson. This process alerts researchers who are sifting through data from billions or even trillions of collisions in search of a rare event, such as the appearance of a Higgs boson.

RE-CREATING THE EARLY UNIVERSE

With its enormous power and capacity, LHC can help physicists study processes that occur under extreme conditions, such as the high-energy environment of the early universe.

Large Hadron Collider

"In the quest to unravel the universe's inner workings," reported Adrian Cho in *Science* in the September 5, 2008, issue, "the 10 September start-up of the LHC marks the beginning of a new age of exploration." Ranked by *Time* as number 5 on the list of best inventions of 2008, the LHC is located in an underground tunnel at the border between France and Switzerland, near Geneva, and has a circumference of 16.6 miles (26.7 km). Two beams of hadrons—protons or lead ions—race around in opposite directions inside the chamber, attaining a maximum energy of 7 TeV. When the particles collide head-on, the total energy is 14 TeV. This energy is several times what the Tevatron, the previously largest particle accelerator, can accomplish.

LHC is one of the most complex pieces of scientific equipment ever devised. The particles must travel in a vacuum to avoid undesired collisions that would scatter the beam, so the track must be almost completely empty—LHC's internal air pressure is even less than on the surface of the Moon. A total of 9,300 magnets sit inside the machine, including 1,624 superconducting main magnets to direct the beam. Since superconductors require extremely low temperatures to operate, LHC employs a refrigeration system that keeps the superconductors at a frigid temperature of -456°F (-271°C)—only a few degrees above absolute zero. Electromagnetic operations accelerate the protons until they traverse the main ring 11,245 times per second, traveling at 99.99 percent of the speed of light.

LHC cost about $6 billion, but will generate an enormous quantity of new data at energies unattainable in previous machines. And with about 600 million proton collisions every second, even rare events should be observable. The amount of data is approximately equal to the content of a few million DVDs every year.

(continues)

(continued)

In the September 5, 2008, issue, Cho quoted Gordon Kane, a University of Michigan physicist, as saying, "The LHC is a 'why' machine." This is because while previous accelerators have revealed the fundamental particles and forces, LHC is powerful enough to answer the more profound question—why these particles and forces exist in the first place. One of the biggest clues will come if and when LHC physicists discover the Higgs boson.

Scientists believe that the universe began with an explosion—the big bang—about 14 billion years ago. Several observations form the basis for this hypothesis, including the expansion of the universe and the presence of a uniform background radiation of a certain energy, both of which suggest that the universe was much smaller in the past than at present. The instant after the big bang, the very early universe was extremely hot and dense. In this high-energy state, even massive particles roamed freely, and there was only one force, which was a unification of the electromagnetic, strong, weak, and gravitational forces. But the temperature fell as the universe expanded, and the energy density dropped. The four forces separated and became distinguishable. Protons and neutrons formed, then nuclei and atoms.

Despite the doomsday warnings, LHC does not have enough energy to replicate the big bang. What it can do, however, is create a high-energy situation that briefly, and in a small volume of space, mimics the conditions of the early universe. The sidebar on page 55 provides more details on this amazing machine.

At the tremendous energies generated in proton-proton collisions, LHC physicists hope to catch a glimpse of an environment similar to the birth of the universe. Although quarks are firmly bound in today's universe, at its earliest stages the universe was too hot and held too much energy for quarks to settle down. The early universe consisted of a jumble of quarks and gluons, the carriers of the strong force. By replicating some semblance of this state of matter, scientists should gain insight into this early phase of the universe and how and why it proceeded to evolve into its present state. Such knowledge would enable researchers to gain a deeper understanding of the fundamental laws of physics by answering the question of why these laws exist and why they have the form and structure that they do. This is one of the most profound mysteries of science.

But the path that science takes is often a tortuous one, full of surprises—no one knows what LHC experiments may uncover. These experiments may add further evidence for the standard model and support what physicists currently believe about the beginning of the universe or the findings may cause theorists to scratch their heads, forcing them to look for new theories.

The path to scientific knowledge is also generally a bumpy one, and LHC has already encountered a setback: On September 19, 2008, just nine days after getting switched on for the first time, some of the superconducting magnets overheated. As Adrian Cho reported in the September 26, 2008, issue of *Science*: "When physicists first sent particles racing through the world's biggest atom smasher on 10 September, the Large Hadron Collider (LHC) at the European particle physics laboratory, CERN, near Geneva, Switzerland, the gargantuan machine purred like a kitten. But only 9 days later, the LHC proved it can also be a temperamental tiger, damaging itself so severely that it will be out of action until next spring." Damage to at least one of the magnets and part of the refrigeration system required millions of dollars to repair and pushed back the accelerator's return to action until late 2009.

CONCLUSION

Despite the setback, LHC will certainly advance particle physics in numerous ways. Some of these advances may come as no surprise, such as finding the highly anticipated Higgs boson, but other findings could be unexpected. Some advances might even be unrelated to physics—one of the most useful and widespread consequences of the use of particle accelerators was the development of the Web.

The Web arose as a solution to the problems researchers at CERN were having with data. Two of the biggest headaches for physicists were the enormous size of the data sets to be analyzed and the huge number of researchers involved in each project. Physicists who work on CERN projects come from universities or research institutes from all over the world and since the researchers do not stay full time at CERN they need remote access to the data. In 1990, the British computer scientist Timothy Berners-Lee developed a data-sharing network in which anyone operating a computer linked to the network could send and receive files. This network gave researchers working from anywhere in the world access to the data, and the system gradually developed into the World Wide Web (www). Particle accelerator laboratories built the earliest Web sites—the first World Wide Web server in the United States, for example, was at SLAC, which came online on December 12, 1991.

Advances in particle physics may be equally important and unanticipated. One possibility is that LHC may have enough power to create tiny black holes, which have caused some people to worry about doomsday scenarios in which the black hole swallows Earth. The strong gravitational field of black holes does attract matter, which gets pulled in and cannot break free, but any black holes in LHC's apparatus will not survive for more than a vanishingly small fraction of a second.

The British physicist Stephen Hawking theorized in the 1970s that tiny black holes evaporate before they have time to drag in much matter. When they evaporate, the black holes emit radiation of a certain type and amount, the detection of which would support the theory—and give physicists a valuable tool to study black holes and their relation to matter and forces such as gravitation. Such studies would be particularly important because particle physicists have been unable as yet to make much headway studying gravitation, since it is so weak compared to the other forces.

Other frontiers of physics that LHC may enlighten include the notion of extra *dimensions.* String theory, the subject of chapter 6, is a speculative theory in physics in which matter consists of tiny strings. These strings vibrate in certain ways, which give rise to particles such as electrons and protons. String theory uses advanced mathematics and is an elegant formulation; the theory excites physicists because it may permit them to explain in mathematic terms the fundamental properties of matter. Among the properties posited in string theory is the existence of dimensions beyond the three dimensions of space and one of time. These extra dimensions, which may number six or seven, provide some wiggle room in which the strings maneuver, explaining some of their properties.

Physicists have discovered little experimental evidence that would support string theory, but the energy necessary to probe these tiny objects has been far out of reach. Although no one is sure if LHC is powerful enough, the world's largest particle accelerator may be able to allow scientists to study these objects, if they exist.

But studying something that moves in extra-dimensional space will not be simple. Particles such as quarks leave their unique signatures in the form of a certain set of particles or radiation, but the traces of strings, should they be real, are difficult to visualize. One possibility is that at the high energies generated during LHC experiments, some objects may hop through these extra dimensions and temporarily vanish! Such unexplained disappearances would offer tantalizing clues to the nature of matter.

Whatever LHC physicists may uncover, particle accelerator experiments will continue to help answer some of the most profound questions at the frontiers of physics. Scientists and students who wish to learn more about the fundamental nature of matter and how the universe evolved are especially interested in accelerators and the experimental opportunities they uniquely provide.

CHRONOLOGY

1897   The British physicist Sir Joseph John Thomson (1856–1940) discovers the electron.

1905   The German-American physicist Albert Einstein (1879–1955) publishes the equation $E = mc^2$, showing how energy, E, is related to mass, m, and the square of the speed of light, c.

1908   The German physicist Hans Geiger (1882–1945) and the New Zealand-British physicist Ernest Rutherford (1871–1937) develop the Geiger counter, an instrument to measure certain types of radiation.

1911   Rutherford and his colleagues discover the atomic nucleus.

The Scottish physicist Charles Wilson (1869–1959) invents the cloud chamber, a method of detecting ionizing radiation that is subsequently used in early particle experiments.

1928   While studying equations of quantum mechanics, the British theoretical physicist Paul Dirac (1902–84) discovered a formula suggesting the existence of a positron, a particle of antimatter.

1929   The American physicist Robert Van de Graaff (1901–67) builds a generator capable of exceptionally high voltages.

1930   The American physicist Ernest O. Lawrence (1901–58) builds the first cyclotron.

1932   The British physicist Sir John Cockcroft (1897–1967) and the Irish physicist Ernest Walton (1903–95) design a high-voltage particle accelerator at Cambridge University in the United Kingdom.

The American physicist Carl Anderson (1905–91) discovers the positron in tracks made in a cloud chamber.

The British physicist James Chadwick (1891–1974) discovers the neutron.

1952   The American physicist Donald Glaser (1926–   ) invents the bubble chamber, a method of detecting particles that is subsequently used in particle experiments.

1954   Construction begins on the first CERN particle accelerator.

1962   The Stanford Linear Accelerator Laboratory, later called SLAC National Accelerator Laboratory, is founded.

1963   Murray Gell-Mann and George Zweig independently propose the existence of quarks.

1967   The National Accelerator Laboratory, later renamed Fermi National Accelerator Laboratory (Fermilab), is founded.

1967–73	Richard Taylor, Jerome Friedman, Henry Kendall, and their colleagues use SLAC to discover the structure of protons, providing evidence for the existence of quarks.
1983	Fermilab's Tevatron, the world's largest accelerator until the Large Hadron Collider is built, begins operation.
	The CERN physicist Carlo Rubbia and his colleagues find the W and Z bosons.
1990	The British computer scientist Timothy Berners-Lee develops a data-sharing network that eventually grows into the World Wide Web.
1994	CERN gets funding approval to build the Large Hadron Collider (LHC).
1995	Two teams of researchers, each numbering about 450, announce that they found the top quark during Tevatron experiments. This quark, the sixth to be found, was the only remaining quark theorized to exist but not yet discovered.
	Walter Oelert and his colleagues at CERN produce atoms of anti-hydrogen, made of a positron and antiproton.
1998	Engineers begin work on the gigantic ATLAS detector for LHC.
2000	The first LHC ring segments arrive at CERN.
2008	LHC begins operation.
	Fermilab researchers place lower limit on Higgs boson's mass—170 GeV/c^2.
	Nine days after start-up, LHC suffers a malfunction that requires more than a year to fix.

FURTHER RESOURCES

Print and Internet

American Institute of Physics. "Early Particle Accelerators." Available online. URL: http://www.aip.org/history/lawrence/epa.htm. Accessed June 22, 2009. This article looks back on the early equipment of the 1920s and 1930s, such as the Cockcroft-Walton accelerator, the Van de Graaff generator, and the cyclotron.

Cho, Adrian. "After Spectacular Start, the LHC injures itself." *Science* 321 (9/26/08): 1,753. This short news article describes the damage sustained during the September 19 accident.

———. "The Overture Begins." *Science* 321 (9/5/08): 1,287–1,289. This news story describes the Large Hadron Collider and its initial operations.

Clery, Dan, and Adrian Cho. "Large Hadron Collider: Is the LHC a Doomsday Machine?" *Science* 321 (9/5/08): 1,291. Journalists summarize the fears about LHC.

Close, Frank. *Particle Physics: A Very Short Introduction.* Oxford: Oxford University Press, 2004. For those who want to cut to the chase, this book offers a concise guide to particles, antimatter, forces, accelerators, and detectors.

Exploratorium Museum. "The Heart of the Matter." Available online. URL: http://www.exploratorium.edu/origins/cern/. Accessed June 22, 2009. An online exhibit of the Exploratorium, a popular museum in San Francisco, provides a guided tour inside CERN and its research facilities. Sections of the exhibit include information on the international community at CERN, the particle accelerators, the experiments, and the experimenters.

Freudenrich, Craig. "How Atom Smashers Work." Available online. URL: http://science.howstuffworks.com/atom-smasher.htm. Accessed June 22, 2009. This article, posted on the howstuffworks Web site, describes the physics of "smashing atoms" and how accelerators work.

Iowa State University. "Iowa State Physicists Part of Research Team Testing Nobel-Winning Theory." Available online. URL: http://www.public.iastate.edu/~nscentral/news/2008/nov/asymmetry.shtml. Accessed June 22, 2009. A team of researchers probes certain particles to find differences in matter and antimatter.

Kirkland, Kyle. *Particles and the Universe.* New York: Facts On File, 2007. Aimed at grades six–12, this book describes the physics of Einstein's theories, quantum mechanics, particle accelerators, nuclear reactions, and cosmology.

Lederman, Leon, and Dick Teresi. *The God Particle: If the Universe Is the Answer, What Is the Question?* New York: Dell Publishing, 1993. Lederman is an eloquent physicist and former director of Fermilab. In this book he and his coauthor describe particle physics and the search for the Higgs boson—the "God particle"—in a witty and entertaining manner.

Live from CERN. "Antimatter: Mirror of the Universe." Available online. URL:http://livefromcern.web.cern.ch/livefromcern/antimatter/. Accessed June 22, 2009. Composed of a series of articles, videos, and webcasts, this Web resource explains the fundamentals and history of antimatter, CERN's experiments with antihydrogen, and much more.

Particle Data Group at the Lawrence Berkeley National Laboratory. "The Particle Adventure." Available online. URL: http://www.particleadventure.org/. Accessed June 22, 2009. This excellent Web resource for students offers an interactive tour of particle accelerators and detectors, the fundamental particles, and many other associated topics, such as antimatter.

Schwarz, Cindy. *A Tour of the Subatomic Zoo: A Guide to Particle Physics,* 2nd ed. New York: Springer Verlag, 1997. This short and accessible book takes the reader step-by-step through the physics of atoms, forces, particles, quarks, the standard model, accelerators, and detectors.

University of Oxford Physics Department. "Inside the Proton." Available online. URL: http://www.physics.ox.ac.uk/documents/pUS/dIS/. Accessed June 22, 2009. This tutorial explains the experiments involving "deep elastic scattering" that particle physics use to probe the structure of protons. Included at this site is a page describing the SLAC experiments that discovered quarks.

Veltman, Martinus. *Facts and Mysteries in Elementary Particle Physics.* Singapore: World Scientific Publishing, 2003. This book offers a complete overview of the subject, including the basics of quantum mechanics and particle theory, and would be a good choice for students with a serious interest in particle physics.

Web Sites

European Organization for Nuclear Research (CERN). Available online. URL: http://public.web.cern.ch/public/. Accessed June 22, 2009. The Web site of CERN contains a wealth of news about the laboratory, especially the Large Hadron Collider, as well as tutorials and articles on the principles and techniques of particle physics.

Fermi National Accelerator Laboratory. Available online. URL: http://www.fnal.gov/. Accessed June 22, 2009. Fermilab's Web site provides a ton of information about particle physics, including the latest research at Fermilab as well as other major accelerator laboratories.

SLAC National Accelerator Laboratory. Available online. URL: http://www.slac.stanford.edu/. Accessed June 22, 2009. News and information on this active research center and about particle physics in general can be found here.

3

Neutrinos—Elusive Particles and the Mysteries of Astrophysics

Modern physics often requires scientists to examine fundamental concepts. Time, for example, seems to be the same for everyone, but the German-American physicist Albert Einstein discovered that the time of an event depends on the observer's coordinate system—the axes and numbers by which the observer specifies the location and time of events—an idea Einstein incorporated into his special theory of relativity. The existence of tiny particles and radiation beyond the range of human senses also forced scientists to concentrate on their methods of detection and observation. One of the most elusive particles is the neutrino.

The act of seeing an object would seem to pose few difficulties. People can see large objects with the unaided eye, and smaller objects are detectable when placed under a magnifying instrument such as a microscope. Yet the process of detecting objects is not as simple as it would appear to be. Although it is not always obvious, the act of detection needs some sort of interaction.

In vision, for example, electromagnetic radiation—light—mediates the interaction. People see an object because it emits light or reflects some of the light coming from a luminous source, such as a lightbulb or the Sun.

The Sun is an important source of neutrinos. This image, taken with the extreme ultraviolet imaging telescope, shows the Sun with a remarkable handle-shaped prominence at the two-o'clock position. *(Extreme Ultraviolet Imaging Telescope Consortium/NASA)*

The light enters the eye and causes a cascade of biochemical reactions that underlie vision. For small objects, the light needs to be magnified because so little of it reflects from the object. Some objects are so small that they do not reflect any electromagnetic radiation in the range of visible light, in which case researchers use lower wavelength radiation, such as X-rays, that the object will scatter or affect in some way. Particle detectors in accelerator experiments sense the passage of tiny objects with a method such as measuring the electric current created when the object ionizes atoms and molecules.

If an object does not easily reflect light, scatter X-rays, ionize atoms, or otherwise interact with other objects, it is exceptionally difficult to detect. Such an object is ghostlike, present but almost undetectable. Even enormous numbers of such objects are practically invisible. This is the case with neutrinos.

A particle that engages in few interactions with other matter might seem to be an uninteresting particle. But quite the contrary is true for the neutrino. Created during important events such as the nuclear reactions in the Sun, these abundant but elusive particles offer a glimpse into the fundamental nature of matter. The properties of neutrinos have an impact on many fields of physics and astronomy, and the study of neutrinos has resulted in a significant revision in the way physicists think about matter and energy. But some of the properties of neutrinos remain mysterious—neutrinos are notoriously difficult to study—and physicists are eager to learn more. This chapter describes what is known and what more there is to learn.

INTRODUCTION

Neutrinos are so difficult to detect that they probably would not have been found if their existence had not been predicted in 1930. And the prediction would not have been made if physicists had not been upset about the peculiar behavior of a radioactive event called beta decay.

In beta decay, a nucleus spontaneously emits a particle that early 20th-century physicists called a beta particle. (Beta [β], the second letter of the Greek alphabet, was an appropriate name, since another type of emission had been called an alpha particle. Alpha is the first letter of the Greek alphabet.) The beta particle is actually an electron, the generation of which occurs when a neutron in the radioactive nucleus converts into a proton, which stays in the nucleus, and an electron, which is emitted. There is another particle involved in this conversion, but researchers initially had no idea of its existence.

However, physicists who studied beta decay found something puzzling—the energy before and after the event did not always seem to add up. When researchers measured the energy of the emitted electrons, the values were variable and did not appear to obey the law of energy conservation. This important law of physics says that energy is not created or destroyed in any process, although it can be converted from one form, such as chemical energy, into another, such as kinetic energy (the energy of motion). Although people often say that they "consume" energy, what they mean in terms of physics is that energy is transformed; for instance, an automobile engine converts the energy contained in the chemical bonds of gasoline into heat and then into the kinetic energy. Physicists have found that in any process, the total amount of energy, when all its forms are included, remains the same.

The problem was that beta decay appeared to contradict this law—the emitted electron did not have enough energy to balance the equation. Physicists wondered if perhaps the electron was losing some of its energy by some process shortly after emission, but the British physicist Sir Charles D. Ellis (1895–1980) and his colleagues conducted careful measurements in the 1920s and reported that this was not true. Some of the energy was mysteriously vanishing. Beta decay also violated another conservation law—that of angular momentum.

Some physicists advocated abandoning the conservation laws. Among them was the Danish physicist Niels Bohr (1885–1962), who substantially contributed to the development of quantum mechanics. In the 2004 book *Are There Really Neutrinos,* Allan Franklin writes, "Bohr made his speculations public in his Faraday lecture to the British Chemical Society on May 8, 1930. Noting the problem . . . he remarked, 'At the present stage of atomic theory . . . we may say that we have no argument, either empirical or theoretical, for upholding the energy [conservation] principle in the case of β-decay disintegrations, and are even led to complications and difficulties in trying to do so.'"

But some physicists believed the "complications and difficulties" were worth the effort, for the conservation laws were important foundations of physics and had proven their value repeatedly, especially in the study of nuclear reactions, as described in chapter 1. The Austrian-American physicist Wolfgang Pauli (1900–58), who also made vital contributions to quantum mechanics, disagreed with Bohr. As noted in George L. Trigg's 1975 book *Landmark Experiments in Twentieth Century Physics,* ". . . in December 1930, Wolfgang Pauli made another suggestion. In a letter to some workers in radioactivity attending a meeting in Tübingen [Germany], he wrote, '. . . I have hit on a desperate remedy to save the laws of conservation. . . . This is the possibility that electrically neutral particles exist. . . .'"

At the time of Pauli's suggestion, neutrons had not yet been found, which is why Pauli suggested the existence of "electrically neutral particles." But unlike the neutron, which was discovered in 1932, the little particles postulated by Pauli served one purpose—to save the conservation law. With no evidence of their existence, and no theoretical justification other than the rescue of a cherished principle, Pauli was going out on a limb. As Trigg writes in his book, "Pauli repeated his proposal in a talk given by invitation at a meeting of the American Physical Society in June 1931; but he never published it, perhaps feeling that the idea should not be taken that seriously."

Some physicists took it seriously, however. In 1934, the Italian-American physicist Enrico Fermi (1901–54) proposed a theory of radioactivity that included Pauli's hypothetical particles, which Fermi called neutrinos—Italian for "little neutral ones." But since neutrinos apparently did not interact very often with other particles, they seemed impossible to detect, and their existence remained theoretical for more than 25 years after Pauli's suggestion.

A breakthrough came in 1956, when the American physicists Clyde Cowan (1919–74), Frederick Reines (1918–98), and their colleagues found evidence of reactions involving neutrinos. In order to maximize their chances of detecting one, the researchers conducted their experiments at the source of a huge quantity of neutrinos—a nuclear reactor. At the Savannah River nuclear reactor in South Carolina, Cowan and Reines used a detector composed of a large volume of water containing chemicals that displayed the products of these rare neutrino reactions.

In the standard model of particles and interactions (see chapter 2), physicists classify neutrinos as leptons. Neutrinos are believed to be elementary particles—they are not composed of any other particles. In 1962, the American physicist Leon Lederman (1922–) and his colleagues found another kind of neutrino. There are a total of three neutrinos known today, each of which is associated with one of the charged leptons—there is an *electron neutrino,* a *muon neutrino,* and a *tau neutrino.* According to the standard model, these are the only neutrinos that should exist. As with all other particles, antiparticles for these neutrinos also exist.

Despite these discoveries, physicists have encountered serious obstacles in studying neutrinos. Theoretical calculations in particle physics, which have been supported by many experiments with particle accelerators, indicated that enormous quantities of neutrinos are created in events such as beta decay and other reactions. These particles are not scarce, in other words. The chief difficulty in their study is detecting a particle that has very little to do with anything else in the universe.

CATCHING AN ELUSIVE PARTICLE

In "Neutrinos Matter," an educational pamphlet funded by the National Science Foundation, the science writer Sharon Butler and the neutrino physicist Janet Conrad noted that scientists can only observe particles when they interact with something. "But neutrinos are the loners of the universe: they rarely interact with each other or anything else. They rip across the great expanse of the universe unperturbed, sailing right through our bodies, on through the crust of the Earth, and out the other side. Neutrinos can happily pass through a wall of lead several hundred light-years thick. In nature, neutrinos bump into other particles only once in a blue moon."

Cowan, Reines, and their colleagues set up their apparatus where Fermi's theory told them that a huge number of neutrinos were flying around. Physicists today are interested not just in neutrinos coming from nuclear reactors on Earth, but neutrinos from the Sun and else-

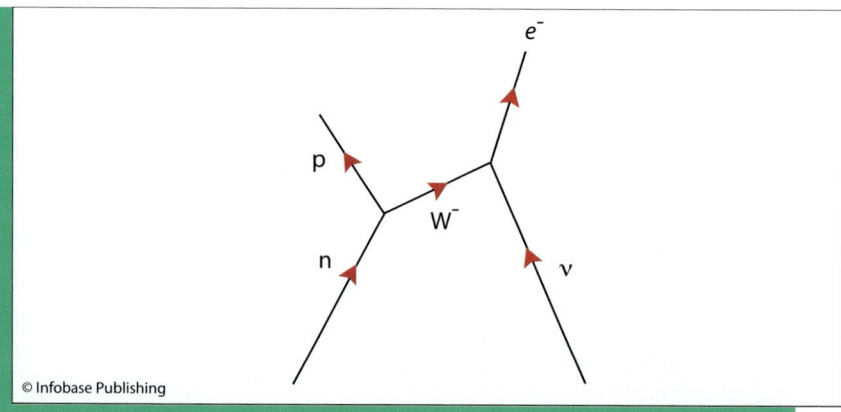

© Infobase Publishing

This diagram illustrates a neutrino v (bottom right), interacting with a neutron, n (bottom left) by exchanging a boson, W^-. The result of the interaction is a proton, p, and an electron, e^-.

where. To increase the number of neutrinos they can observe, researchers build huge detectors—the greater the size, the greater the chance for an interaction that occurs with exceptional rarity.

Technicians cleaning some of the photomultipliers in Super-Kamiokande *(Kamioka Observatory, ICRR, University of Tokyo)*

Čerenkov Radiation

The speed of light in a vacuum—186,000 miles/second (300,000 km/s)—is nature's speed limit. Not even particle accelerators can accelerate anything beyond this speed. But light, and other electromagnetic radiation, slows down when it travels through matter because it interacts with its components or particles. For example, the speed of light in water is about 140,000 miles/second (225,000 km/s). A fast-moving particle can exceed this speed in water (at least briefly, until it crashes into a water molecule), so particles can travel faster than the speed of light in certain media such as water.

But when a charged particle exceeds the speed of light in water or some other medium, it emits electromagnetic radiation. The Russian scientist Pavel Alekseyevich Čerenkov (1904–90) was the first to study this phenomenon, so the radiation is called Čerenkov radiation in his honor. This radiation appears bluish. The emitters are often small charged particles such as electrons, since they are tiny enough to zip through various media at a tremendous speed if they have enough energy.

Why does Čerenkov radiation occur? The reason is similar to the production of sonic booms—loud noises created when an airplane exceeds the speed of sound. When an airplane goes faster than sound, the noises it generates build up into a shock wave. In the case of Čerenkov radiation, a particle traveling faster than electromagnetic radiation in a certain media emits a "shock wave" of light.

One of the most impressive neutrino detectors is the Super-Kamiokande (Super-K), located about 3,280 feet (1,000 m) below ground in the Kamioka Mozumi mine in Japan. (An older detector was called Kamiokande, named for the mine and the initials for "nucleon decay experiment." In 1996, researchers finished an upgrade—the Super-Kamiokande detector.) The detector consists of a huge tank, 136 feet (41.4 m) by 129 feet (39.3 m), which holds about 12,500,000 gallons (47,318,000 L) of pure water. Researchers placed Super-K far underground to shield the detector from stray radioactive particles that they did not want to study. Neutrinos can travel through a lot of rock without being stopped, so the ground filters out the unwanted particles without affecting the particles that researchers want to measure.

As in other particle detectors described in chapter 2, Super-K looks for signs left by a passing neutrino, only a few of which interact with the water. When a neutrino bumps into an electron or the nucleus of a water molecule, it interacts by exchanging a particle known as a boson, as illustrated in the figure on page 70. In the case of electron neutrinos, the interaction results in a free electron; in the case of muon neutrinos and tau neutrinos, the result is a muon and tau particle, respectively. The process generates a charged particle with enough energy to travel through water faster than light travels through water. As described in the sidebar on page 72, such a high-speed particle emits electromagnetic radiation known as Čerenkov radiation.

The wall of the tank in Super-K houses about 11,000 sensitive light detectors called photomultipliers. Electronic circuitry in these instruments amplifies, or multiplies, any electromagnetic radiation that strikes the instrument. These photomultipliers record Čerenkov radiation created during the neutrino interactions. By analyzing the timing and location of this radiation, physicists can reconstruct the path of the neutrino and the direction from which it came. Super-K spots a few hundred neutrinos per month.

SOLAR NEUTRINOS

The Sun produces many neutrinos during its nuclear reactions. Since neutrinos fly through most material, including the layers of the Sun, these particles offer an insight into the inner workings of solar processes. In the 1960s, the American researchers Raymond Davis, Jr. (1914–2006), John N. Bahcall (1934–2005), and their colleagues calculated the number of neutrinos the Sun should theoretically be producing. All of these neutrinos were electron neutrinos. Sophisticated detectors such as Super-K had not yet been built, so the researchers set up a tank about the size of a swimming pool in the Homestake Gold Mine in South Dakota and filled it with chlorine-based cleaning fluid. They analyzed the number of passing electron neutrinos by counting the radioactive atoms left in their wake. The researchers expected to find about one neutrino a day. In 1968, they announced their first result, and the number was much smaller than expected—the detector observed a neutrino about every three or four days.

Bahcall, in the article "Solving the Mystery of the Missing Neutrinos" published in 2004 on the Nobel Foundation's Web site, discussed how scientists approached this discrepancy. "Three classes of explanation were suggested to solve the mystery. First, perhaps the theoretical calculations were wrong." The error could have been with the prediction or the means by which the neutrinos were counted in the detector. "Second, perhaps Ray's [Raymond Davis's] experiment was wrong. Third, and this was the most daring and least discussed possibility, maybe physicists did not understand how neutrinos behave when they travel astronomical distances."

When faced with a number of plausible scenarios, scientists attempt to find the correct one by eliminating all the rest. In the case of the missing neutrinos, physicists checked and rechecked the theoretical calculations, until they were convinced that no error had been made. Davis and his colleagues also repeated the experiments in a variety of ways, but the discrepancy between prediction and experiment continued. Bahcall wrote, "What about the third possible explanation, new physics? Already in 1969, Bruno Pontecorvo (1913–93) and Vladimir Gribov (1930–97) of the Soviet Union proposed the third explanation listed above, namely, that neutrinos behave differently than physicists had assumed. Very few physicists took the idea seriously at the time it was first proposed, but the evidence favoring this possibility increased with time."

A team of Japanese and American researchers, led by Masatoshi Koshiba (1926–   ) and Yoji Totsuka (1942–2008), used the Kamiokande neutrino detector to confirm that the Sun emitted fewer neutrinos than nuclear and particle physics predicted. This result set the stage for the Super-K. The solution to the missing solar neutrinos involved, as Bahcall noted, "new physics."

NEUTRINO OSCILLATION—CHANGING FROM ONE TYPE TO ANOTHER

One way to account for the missing neutrinos was to assume they had changed form on the journey from the Sun to the neutrino detectors on Earth. Recall that there are three different types or "flavors" of neutrino, each associated with one of three fundamental particles—electron, muon, and tau. The missing solar neutrinos were electron neutrinos. If some of the electron neutrinos somehow had changed into another kind of neutrino before reaching the detector, this would explain why researchers found fewer than expected—they were looking for electron neutrinos, not the other kinds. This was the hypothesis that Pontecorvo and Gribov proposed. The problem would be solved if neutrinos switched, or oscillated, from one flavor to another.

But this hypothesis clashed with what physicists believed they knew about neutrinos. Recall that neutrinos had offered a solution to the missing energy, required by the law of energy conservation, in certain types of radioactive decays. Neutrinos nicely filled this role assuming they were electrically neutral and without mass. (Particles with no mass may sound strange, but there are examples of such particles—photons, the particles of light, have no mass. Note that this means the particles have no "rest" mass; they do have momentum and energy.) The possibility that neutrinos may be changing state was not a problem, since quantum mechanics, the theory that explains particle behavior, describes such changes. What did trouble physicists was that the theory required particles to have mass in order to undergo these changes.

On June 5, 1998, Super-K researchers, led by Yoji Totsuka, announced evidence for neutrino oscillation—and therefore evidence for neutrino mass. In the press release, the researchers wrote, "Reflecting on the significance of the new finding, we note that massive neutrinos must now be incorporated into the theoretical models of the structure of matter and that astrophysicists concerned with finding the 'missing or dark matter' in the universe, must now consider the neutrino as a serious candidate." This missing matter is needed to explain discrepancies in gravitational forces that astronomers have found in galaxies in the universe. Further astronomical implications of neutrinos are discussed in the following two sections.

Super-K researchers did not study solar neutrinos in this particular experiment, but instead examined neutrinos created in the upper atmosphere. These neutrinos are produced during interactions as high-speed particles called cosmic rays bombard Earth. (Sources of these high-speed particles include the Sun and other bodies throughout the universe.) All three types of neutrinos are generated during these interactions.

To see if neutrinos may be oscillating, researchers examined the number of neutrinos of a given type coming from opposite directions.

The rationale for this experiment was that these neutrinos traveled different distances. Neutrinos coming from overhead—near the surface—had traveled only a short distance from the atmosphere to the underground detector. But neutrinos coming from below had been created on the other side of the world and had traveled all the way through the Earth. (Which, since neutrinos rarely interact with other matter, is not surprising.) Earth's diameter is about 8,000 miles (12,900 km), and in this extra distance some of the neutrinos coming from below had a little extra time to change states. Super-K researchers measured the ratio of neutrinos coming from above and below. Since there was no difference in the conditions above and below the detector, the same number of neutrinos of a given type should be created. If the data indicated more of a certain type in one of the directions, this would support the notion that neutrinos were oscillating.

The journalist Dennis Normile reported the results in the June 12, 1998, issue of *Science.* "For electron neutrinos, Super-Kamiokande caught equal numbers going up and coming down. But for muon neutrinos there was a big difference. In 535 days of operations, Super-Kamiokande counted 256 downward muon neutrinos and just 139 upward ones. The large number of observed neutrinos and the magnitude of the difference reduce the chances of the finding being a statistical fluke, say team members. Taken together, the data indicate that the muon neutrinos are oscillating, perhaps to tau neutrinos, which the detector cannot pick up." Normile quoted Sheldon Glashow, a physicist at Harvard University, as saying, "It is one of the most important discoveries in particle physics of the last few decades."

Further support for this dramatic finding came on June 18, 2001. This time the results were from an international collaboration of scientists, led by Arthur McDonald (1943–   ), who used the new Sudbury Neutrino Observatory (SNO). This detector, completed in 1999, is located in a mine in Sudbury, Ontario, Canada, at a depth of about 6,800 feet (2,070 m). The tank is 39.4 feet (12 m) in diameter, with almost 10,000 photomultipliers monitoring the "heavy" water (deuterium oxide) contained within.

SNO researchers conducted sophisticated measurements and determined, in conjunction with results from other detectors, the total number of neutrinos coming from the Sun, including all three flavors. This number agreed with the theoretical prediction. Only a third of solar neutrinos were electron neutrinos—the other electron neutrinos had changed en route. Bahcall noted in "Solving the Mystery of the Missing Neutrinos" that the solution had been definitively found. "The smoking gun was discovered. The smoking gun is the difference between the total number of neutrinos and the number of only electron neutrinos. The missing neutrinos were actually present, but in the form of the more difficult to detect muon and tau neutrinos."

Neutrino oscillation required a change in particle physics. The standard model of particle physics had assumed that neutrinos did not have mass, but they must have mass in order to oscillate. Adapting the model to this new finding did not cause any significant problems, but the existence of neutrino mass, considering the abundance of these particles, had astronomical implications. In order to better understand these implications, physicists needed to determine exactly how much mass a neutrino possesses. This measurement requires additional effort, because while the neutrino detector experiments provided evidence that neutrinos have mass, they did not indicate the amount.

DETERMINING THE MASS

Pauli balanced the energy in beta decays with the neutrino, but he assumed it had zero mass. As explained in the sidebar on page 6, Einstein discovered that energy, E, is related to mass, m, by the equation $E = mc^2$, where c is the speed of light in a vacuum. Since neutrinos have mass according to the Super-K and SNO experiments, they add a little more energy than Pauli had anticipated. But their mass cannot be too much, otherwise the law of energy conservation would be wrong in the other direction—neutrinos would account for too much energy.

Weight is the force of gravitation acting on a mass. Individual particles are so tiny that they have negligible weight in terms of units such as ounces or grams, so physicists often describe the mass of a particle in terms of Einstein's equation: $m = E/c^2$. The unit of energy used in this formula is usually the electron volt, eV, which equals the kinetic energy gained by an electron as it accelerates through a potential difference of one volt. (See chapter 2.) For example, the electron has a mass of 511,000 eV/c^2.

If the mass of a neutrino was comparable to an electron, this would pose a serious astronomical dilemma. Gravitational attraction is proportional to the masses of the two interacting bodies, and although the

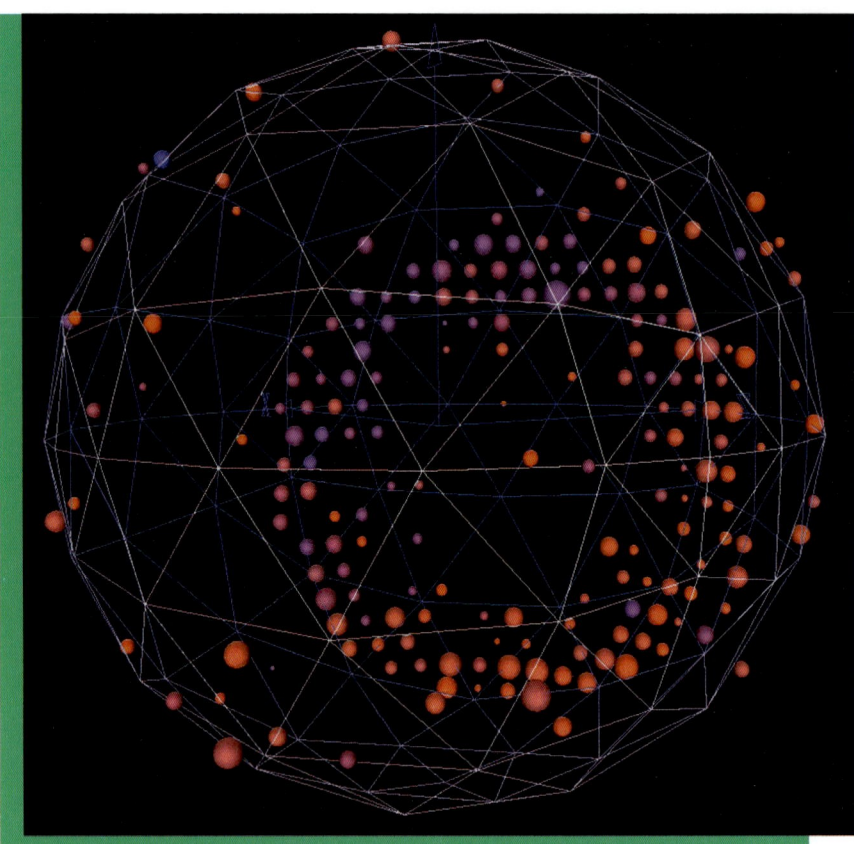

Light registering on sensors in this neutrino detector, part of Fermilab's booster neutrino experiment (BooNE) project, records the collision of a muon neutrino with an atomic nucleus. *(Fermilab Visual Media Services)*

force of gravitation is much weaker than the other three forces—the strong nuclear force, the weak nuclear force, and electromagnetism—it is an important factor for matter in bulk. The force of gravitation is negligible for two particles, but the collective mass of a lot of particles, such as the many particles that constitute Earth, exerts a tremendous force. Neutrinos are common throughout the universe because many arose during the origin of the universe, as well as from the reactions occurring frequently in the Sun and other stars. All of that collective mass would be enough to contract the universe.

But the universe is expanding, and the rate of expansion seems to be accelerating. This expansion puts a limit on neutrino mass. Neutrinos must have a lot less mass than electrons.

How much mass a neutrino has is an important question, but a difficult one to answer. The problem is the same as before—neutrinos engage in few interactions, and a measurement of a property such as mass requires some form of interaction with the measuring apparatus.

Researchers continue to use huge detectors to observe neutrinos from the Sun or cosmic ray interactions, but some scientists have configured a more controllable experimental situation. For example, physicists at Fermi National Accelerator Laboratory (Fermilab) are using their accelerator equipment to generate a large number of neutrinos, and then routing this neutrino beam from the laboratory in Batavia, Illinois, to an underground detector in a mine in Soudan, Minnesota, about 450 miles (725 km) away. (More information on Fermilab can be found in the sidebar on page 43.) This configuration gives researchers much more control over the timing and intensity of neutrino experiments. The experiment is called the Main Injector Neutrino Oscillation Search (MINOS); the main injector is an accelerator at Fermilab that is involved in many different experiments.

Neutrinos do not take long to cover the 450 miles (725 km)—the trip only lasts about 0.0025 second. A detector at Fermilab samples the beam, giving researchers a measurement of the number of particles it contains. The detector in Minnesota makes its observation 0.0025 second later. In that short span of time, some of the neutrinos will have oscillated, or changed state. By comparing the number and types of neutrinos at the Minnesota detector with those at Fermilab, physicists can study neutrino oscillation in a precise manner.

Oscillation rates and properties reveal much about neutrinos and are especially important because they are a reflection of the particle's mass. According to the mathematical equations of quantum mechanics, the probability that a neutrino will change flavor depends on the difference in mass of the flavors. Due to this relationship, a measure of oscillation rate provides information on the difference in the masses of the different neutrino types.

On March 30, 2006, Fermilab issued a press release describing some of the MINOS results. In accordance with earlier observations, MINOS researchers found that neutrinos oscillate and must have mass. MINOS physicists calculated a rate of oscillation that "yields a value of delta m^2, the square of the mass difference between two different types of neutrinos, equal to 0.0031 eV^2." (The press release omitted the speed of light in the units—the value of delta m^2 is more precisely 0.0031 (eV/c^2)^2 or 0.0031 eV^2/c^4.)

The difference in mass between neutrino types appears to be tiny, as one would expect if the neutrino mass is small. Further MINOS experiments, along with similar projects at Fermilab and elsewhere, will continue to explore this issue.

NEUTRINO ASTRONOMY

Astronomers as well as physicists are interested in neutrinos. These elusive particles have drawn astronomers' attention for several reasons. One of the most important reasons is the effect that neutrino mass has on cosmology—the study of the universe.

Observations of radiation coming from galaxies as well as other sources show that the universe has been expanding since its creation, some 14 billion years ago, in an explosion known as the big bang. The amount of matter in the universe influences the rate of expansion and whether it will continue or reverse one day, if the force of gravitation is strong enough to begin a phase of contraction. Astronomers have recently discovered that the expansion rate has been accelerating, which leads them to believe that there is an extra push, or force, driving this accelerated expansion. Astronomers have speculated that some form of unknown energy, called *dark energy,* is responsible.

In addition to dark energy, astronomers who study the distribution and motion of galaxies have found that gravitation is much stronger than expected given the amount of visible matter, such as stars and clouds of dust and gas. Since equations describing the force of gravitation have been repeatedly confirmed, astronomers suspect that the discrepancy is due to the presence of unseen, or *dark matter.* According to this hypothesis, the additional mass from dark matter accounts for the increased gravitational force.

Neutrinos may play some role in these hypotheses. Perhaps neutrinos, if they have enough mass, are the constituents of dark matter. If so, they would explain the gravitational excess. And the difficulty in observing neutrinos explains why dark matter is not easy to spot.

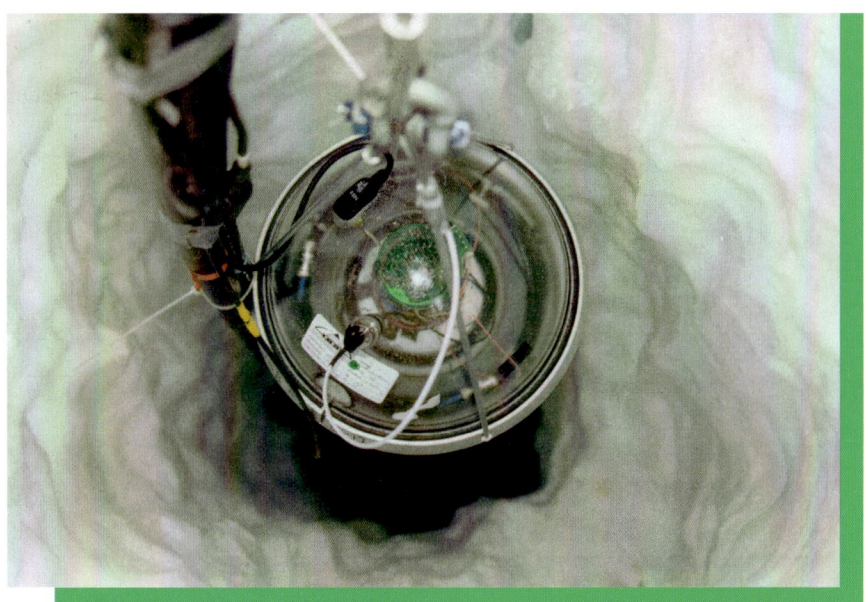

This neutrino telescope, AMANDA, is located in the ice of Antarctica. *(DESY Zeuthen)*

Although neutrinos would seem an ideal candidate for dark matter, measurements and calculations of their mass suggest they are not heavy enough to explain the astronomical observations. Instead, some astronomers have postulated the existence of weakly interacting massive particles (WIMPs). A WIMP is extremely difficult to observe, similar to a neutrino, but is much heavier. These theoretical particles are also reminiscent of neutrinos in that their existence has been postulated in order to explain a puzzling observation, in this case of discrepancies between gravitation and the amount of observable matter.

Another important reason that neutrinos excite astronomers is that they offer a valuable tool to study astronomical objects and events. Observational techniques in astronomy rely on emissions or reflections from distant objects; for instance, astronomers study stars by analyzing the electromagnetic radiation these objects emit. The Sun emits a vast number of neutrinos. Although neutrinos are much more difficult to detect than light, they give astronomers as well as physicists important clues about the nuclear reactions occurring in the Sun. These reactions

Supernova—An Exploding Star

All stars have a limited lifetime. They are born as a swirling cloud of gas contracts and coalesces due to gravitational forces, eventually reaching a pressure and temperature that can sustain nuclear fusion. (See chapter 1.) The energy of these nuclear reactions yields radiation, which counterbalances the force of gravitation. A star maintains this equilibrium for many years, depending on its size—in the case of the Sun, a relatively small star, this phase lasts about 10 billion years or so.

But once most of the atoms have fused and no further energy can be squeezed out of them, gravitation wins. The star begins to contract. For small stars such as the Sun, the contraction tends to stop when matter becomes highly compressed. A small star winds up as a dense and relatively inert sphere of matter called a white dwarf. But with stars that have more mass, the result can be a spectacular explosion—a supernova. Supernovas get their name from *nova,*

take place in the depths of the Sun, and astronomers have few ways to study these hidden regions except with neutrinos that are created in the reactions. Because neutrinos rarely interact with other matter, they fly through the Sun as easily as Earth, providing a glimpse of otherwise obscure processes for astronomers who are able to catch a few neutrinos in detectors such as Super-K and SNO.

Neutrino astronomy is particularly beneficial in the study of a *supernova.* As described in the sidebar above, a supernova is an exploding star. After reaching the end of its nuclear fuel, large stars undergo a cataclysmic explosion, briefly shining as bright as a few billion Suns— about as bright as a small galaxy. These events are rare, occurring only about once or twice a century in the large Milky Way Galaxy, in which the Sun and its planets reside.

Latin for "new," which astronomers used to describe stars that suddenly increase in brightness. With a brightness level up to billions of times that of a normal star, a supernova is certainly an extraordinary nova!

Stars with more mass have greater pressures and temperatures in their center, which means that these stars can fuse heavier nuclei. Fusion of nuclei up to iron (atomic element number 26) releases energy. But fusion of heavier nuclei requires an input of energy, so they do not occur spontaneously in stars. The energy needed to generate nuclei heavier than iron comes from the tremendously energetic events of a supernova.

Astronomers classify supernova events into two types, called Type I and Type II. But no one is certain how these explosions take place, and astronomers continue to study the phenomena, although the rarity of these events limits how often they can be studied. Supernova events are particularly intriguing since they are the "ovens" in which many elements are "cooked"—elements that make up a substantial part of Earth and all the life that exists here.

Many of the details of what happens during a supernova remain mysterious. A series of last-minute nuclear reactions occur, and these processes generate neutrinos. If astronomers can capture a "snapshot" of a supernova by studying the emitted neutrinos, they may be able to gain important insights into the kind of reactions that take place, and the order in which they occur.

In 1987, this idea was tested when a supernova occurred relatively close to Earth. Although the supernova was outside the Milky Way Galaxy, it happened in the Large Magellanic Cloud, a nearby galaxy. In a 1997 issue of *Los Alamos Science,* the researcher Marc Herant and his colleagues summarized the excitement over the role of neutrinos in a supernova. "This pivotal and wondrous function of the neutrino, so much in contrast with its usual marginal position, received triumphant vindication in February 1987, when two underground detectors recorded a burst of neutrinos and a spectacular supernova was later observed by astronomers worldwide. The astrophysical community was elated!" It was the first time neutrinos had been observed from a supernova, although only a few dozen were detected.

Because neutrinos interact so little with matter, they zip through the dense gases in the outer portions of the exploding star much more efficiently than light. This property explains why neutrinos preceded other emissions in the 1987 supernova. (Light is slightly faster than neutrinos, but neutrinos had a head start.) Because neutrinos are the first to arrive, researchers are hoping to monitor these particles to provide an alarm, which is known as the SuperNova Early Warning System (SNEWS). Although neutrinos may beat light by only a few hours, it is enough time to alert researchers to prepare for the event. By training their instruments on future supernovas, astronomers will be able to use telescopes and neutrino detectors to learn more about these astonishing explosions.

CONCLUSION

Although physicists have made much progress since Pauli predicted the existence of neutrinos in 1930, many questions remain. Researchers have yet to pin down the mass of these particles, which is of prime importance to astronomy as well as physics. Scientists also want to use neutrinos to study astronomical objects such as a supernova and other neutrino emitters.

Other questions about neutrinos strike at the most fundamental aspects of particle physics. The standard model consists of three types of neutrinos—electron, muon, and tau—and their antiparticles. Physicists have discovered all three types, and the assumption is that these are all the neutrinos that exist. But this assumption is not proof. An earlier assumption that neutrinos have no mass has been disproven by the Super-K and SNO experiments in 1998 and 2001, respectively, along with several other more recent measurements on neutrino oscillations. Although there is currently no evidence for a fourth neutrino type, there is also no proof that such a particle does not exist. The question is an open one. If other kinds of neutrinos are eventually found, the discovery would lead to a revision (once again) of the standard model—perhaps a profound revision, depending on the nature of the discovery.

The most pressing need in this frontier of science is probably the development of more and improved detectors. Without an enhanced means of observing these elusive particles, progress in neutrino physics will be slow and haphazard.

To meet this goal, physicists are getting ambitious. For example, on May 30, 2008, workers finished building the first underwater neutrino detector, located deep in the Mediterranean Sea off the coast of France. The international project, involving researchers from universities and institutions in

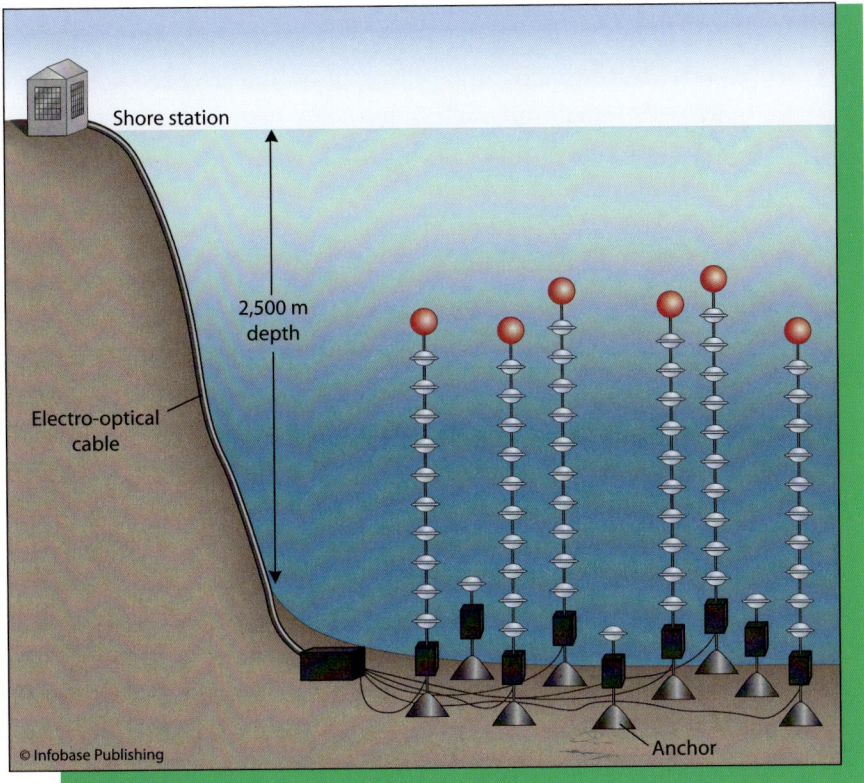

The ANTARES telescope consists of vertical strings of sophisticated electronics anchored to the seabed. Cables carry the data to computer stations on the shore. (Note: figure not drawn to scale.)

France as well as Spain, Italy, Germany, Romania, United Kingdom, the Netherlands, and Russia, is called ANTARES, short for Astronomy with a Neutrino Telescope and Abyss Environmental Research. Antares is also the name of a prominent star in the constellation Scorpius.

As the name of the detector suggests, one of the primary duties of ANTARES is the detection of neutrinos from astronomical sources. But unlike other "telescopes," ANTARES is aimed downward instead of toward the sky. The detector consists of 12 vertical strings of photomultiplier tubes, anchored to the bottom of the sea at a depth of 8,200 feet (2,500 m). Strings reach about 1,300 feet (400 m) above the seabed, and a total of about 1,000 photomultipliers are used. The detector senses Čerenkov radiation, but in this case the fluid of the detector is part of the Mediterranean Sea. Surface area covered by the detector amounts to about 0.04 mile^2 (0.1 km^2). The arrangement, illustrated in the figure, is focused on detecting signatures of neutrinos that are passing through Earth. Only with weakly interacting particles such as neutrinos can a telescope be aimed toward the ground! Cables of about 25 miles (40 km) in length carry signals from the detector to a shore station.

The use of water at the bottom of a sea obviates the need for a specially constructed tank to be placed in a mine or some other location deep underground. But seawater has its disadvantages as well. Although very little sunlight penetrates to the depth of the ANTARES detector, many marine organisms are bioluminescent—they produce their own light with special chemical reactions. This faint light constitutes "noise" that can obscure signals from neutrinos, reducing the detector's efficiency.

Researchers are hoping that ANTARES, and other new detectors, will give them an improved mechanism to observe astronomical sources of neutrinos. Many objects emit neutrinos, especially high-energy objects such as centers of galaxies that appear to be highly active. These objects are among the most distant, which means that electromagnetic radiation from these objects tends to get blocked during its passage through the huge amount of dust and other intervening astronomical bodies. But many neutrinos make the passage relatively unimpeded. ANTARES researchers will try to take advantage of this unique window on the universe.

In his article "Solving the Mystery of the Missing Neutrinos," Bahcall wrote, "I am astonished when I look back on what has been accomplished in the field of solar neutrino research over the past four decades.

Working together, an international community of thousands of physicists, chemists, astronomers, and engineers has shown that counting radioactive atoms in a swimming pool full of cleaning fluid in a deep mine on Earth can tell us important things about the center of the Sun and about the properties of exotic fundamental particles called neutrinos." This frontier of science remains an astonishing and fertile field of research, as scientists use water deep in the Mediterranean Sea to peer across vast regions of space, searching for the secrets of the universe.

CHRONOLOGY

1899	The French physicist Henri Becquerel (1852–1908) investigates the radioactive process known as beta decay.
1927	The British physicist Sir Charles D. Ellis (1895–1980) and his colleagues report that the emitted electrons in beta decay do not account for all of the energy in the decay process.
1930	The Austrian-American physicist Wolfgang Pauli (1900–58) proposes the existence of a small neutral particle to balance the energy equation in beta decay.
1934	The Italian-American physicist Enrico Fermi (1901–54) expounds a theory of radioactivity that includes Pauli's hypothetical particles, which Fermi calls neutrinos.
1956	The American physicists Clyde Cowan (1919–74), Frederick Reines (1918–98), and their colleagues discover the neutrino.
1962	The American physicist Leon Lederman (1922–) and his colleagues show that there is more than one kind of neutrino.

1968	The American researchers Raymond Davis, Jr. (1914–2006), John N. Bahcall (1934–2005), and their colleagues fail to detect as many solar neutrinos as theory predicted.
1969	The Italian-Russian physicist Bruno Pontecorvo (1913–93) and his colleagues propose that neutrinos switch or oscillate between the different types.
1987	The supernova designated 1987A is the first such event in which researchers observe neutrinos.
1996	Researchers finish design and construction of the Super-Kamiokande (Super-K) neutrino detector in Japan.
1998	Super-K researchers, led by Yoji Totsuka, report experimental evidence that neutrinos oscillate, which explains the missing solar neutrinos.
1999	Researchers finish the design and construction of the Sudbury Neutrino Observatory (SNO) in Canada.
2001	SNO researchers confirm neutrino oscillation.
2006	Physicists on the Main Injector Neutrino Oscillation Search (MINOS) discover evidence of muon neutrino oscillation that indicates a tiny mass difference between different types of neutrinos.
2008	Researchers finish design and construction of ANTARES, the first underwater neutrino observatory.

FURTHER RESOURCES

Print and Internet

Bahcall, John N. "Solving the Mystery of the Missing Neutrinos" (4/28/04). Available online. URL: http://nobelprize.org/nobel_prizes/

physics/articles/bahcall/. Accessed June 22, 2009. This accessible account of the solar neutrino mystery and its solution is written by one of the main participants in this research.

Butler, Sharon, and Janet Conrad. "Neutrinos Matter." Available online. URL: http://www-boone.fnal.gov/about/nusmatter/. Accessed June 22, 2009. This interesting article explains the properties of neutrinos and how researchers study them.

Fermi National Accelerator Laboratory. "MINOS Experiment Sheds Light on Mystery of Neutrino Disappearance." Available online. URL: http://www.fnal.gov/pub/presspass/press_releases/minos_3-30-06.html. Accessed June 22, 2009. Fermilab announces the results of a study showing a slight difference in mass between types of neutrinos.

Franklin, Allan. *Are There Really Neutrinos?* Boulder, Colo.: Westview Press, 2004. Franklin gives a thorough account of the history of neutrino research.

Herant, Marc, Stirling A. Colgate, Willy Benz, and Chris Fryer. "Neutrinos and Supernovae." *Los Alamos Science* 25 (1997): 64–79. This article provides a summary of the excitement about using neutrinos to study supernova events.

Mann, Alfred K. *Shadow of a Star: The Neutrino Story of Supernova 1987A.* New York: W. H. Freeman & Company, 1997. Supernova 1987A was the first time researchers observed neutrinos from a supernova. The exciting story is told here.

Normile, Dennis. "Weighing In on Neutrino Mass." *Science* 280 (6/12/98): 1,689–1,690. This news brief discusses the Super-Kamiokande experiment that observed neutrino oscillation.

Super-Kamiokande Collaboration. "Evidence for Massive Neutrinos." News Release (6/5/98). Available online. URL: http://www.phys.hawaii.edu/~jgl/sk_release.html. Accessed June 22, 2009. Super-Kamiokande researchers announced their experimental results supporting neutrino oscillation.

Trigg, George L. *Landmark Experiments in Twentieth Century Physics.* Mineola, N.Y.: Dover Publications, 1995. The experiments discussed in this historical account include the discovery of X-rays, superconductivity, transistors, laser, quantum tunneling, neutrinos, and many others.

Verkindt, Didier. "History of the Neutrinos." Available online. URL: http://wwwlapp.in2p3.fr/neutrinos/aneut.html. Accessed June 22, 2009. This Web resource contains numerous articles on the discovery of neutrinos and the history of neutrino experiments.

Web Sites

ANTARES. Available online. URL: http://antares.in2p3.fr/. Accessed June 22, 2009. The Web site for ANTARES contains news and information on this large, underwater neutrino detector.

Super-Kamiokande. Available online. URL: http://www-sk.icrr.u-tokyo.ac.jp/sk/index_e.html. Accessed June 22, 2009. The Web site for the Super-K detector contains news and information on this important neutrino observatory.

SuperNova Early Warning System (SNEWS). Available online. URL: http://snews.bnl.gov/. Accessed June 22, 2009. SNEWS updates are posted on this Web site, along with background information, mailing lists, information for amateur astronomers, and instructions on how to get a SNEWS alert.

4

SUPERCONDUCTORS— PERFECT ELECTRICAL CONDUCTORS

In many cases, a technological breakthrough gives a tremendous boost to one or more of the frontiers of science. One of the best examples is the discovery of superconductors.

A superconductor is a special electrical conductor. Any electrical conductor permits, or conducts, the flow of electric charges—a current, usually measured in amperes or amps—though conductors generally impede or resist this flow to a varied extent, depending on the material. But superconductors have no resistance, which means they are the most efficient electrical conductors. This property, along with some of their magnetic properties, makes superconductors a vital component in a number of important engineering applications, as well as an interesting phenomenon for physicists to study.

All of the superconductors known today operate only when cooled to an extremely low temperature. At room temperature, say 68°F (20°C), these materials are just ordinary conductors. It was not until the early 20th century, when people developed the technology to attain extremely low temperatures, that researchers could discover superconductivity.

Superconductors are essential for particle accelerators, described in chapter 2, as well as other sophisticated devices, such as certain machines that image brain activity. But the need to keep superconductors cold limits their usefulness. Physicists have been finding materials that retain their superconducting properties at increasingly high temperatures, but no superconductor can yet operate anywhere close to room temperature—which would immensely extend their applicability. Although modern physics can explain low-temperature superconductors, the more recently discovered higher temperature superconductors remain mysterious, and the absence of a theory to explain all forms of superconductivity hurts the search for new ones. This chapter describes these fascinating materials and the search for "warmer" superconductors—and a theory that would explain this phenomenon.

INTRODUCTION

Heat flows from warm objects to cooler ones—this is the "downhill" direction for thermal energy. When heat can flow—by contact between objects or carried by a flow of air or the emission and absorption of radiation—objects will eventually reach the same temperature. This flow of heat means that keeping an object cool in a warm environment is difficult and lowering an object's temperature far beyond its environment requires special equipment.

In addition to the problems of constraining the flow of heat, there is a lower limit to an object's temperature. This temperature is called *absolute zero,* and it is the 0 value in the absolute, or *Kelvin scale* (denoted K), named in honor of Lord Kelvin (Sir William Thomson). Absolute zero registers -459.67°F (-273.15°C), which equals 0 K. (There are no degrees in the absolute scale—the unit is the kelvin.) The reason for this limit is that on an atomic level, heat is motion, and at absolute zero, motion is at its minimum. (Motion is not zero at this temperature because the principles of quantum mechanics stipulate some amount of motion for all objects.) Temperatures of a few degrees above absolute zero are hard to reach—everything else is a lot warmer, and heat tends to flow in to the cooled object, raising its temperature.

Objects undergo phase transitions at certain temperatures, which is important in science as well as the environment. Water, for example, exists on Earth as a vapor, liquid, and solid, depending on the temperature. Other substances experience changes in phase only at drastic temperatures. Helium, for instance, is a gas at a wide range of temperatures and turns into a liquid at extremely cold temperatures. The Dutch physicist Heike Kamerlingh Onnes (1853–1926) was the first to liquefy

Insulated containers of liquid helium *(Cornell—LEPP)*

helium in 1908 after he found the means to cool helium to the temperature of -452°F (-268.9°C). George Trigg, in his book *Landmark Experiments in Twentieth Century Physics,* explains, "When a gas is allowed to expand through an orifice from a region of high pressure to one of lower pressure, its temperature decreases, provided it is initially already below a critical value, the inversion temperature, that depends on the gas. To obtain liquefaction of any significant fraction of the gas, it must initially be below about one-third of the inversion temperature. For helium, the inversion temperature is 51 K [-368°F (-222°C)]; it is this fact that makes necessary all the preliminary cooling stages, and explains why the very achievement of liquefaction was a major feat."

After Onnes's work, scientists could use these procedures, as well as liquid helium itself, to generate temperatures that approach absolute zero. Onnes was a beneficiary of his own technological breakthrough—which is not always the case in science—when he discovered superconductivity in 1911, only three years after he achieved the liquefaction of helium.

Researchers were already aware that electrical resistance depends on temperature. But Kamerlingh Onnes found that below a certain

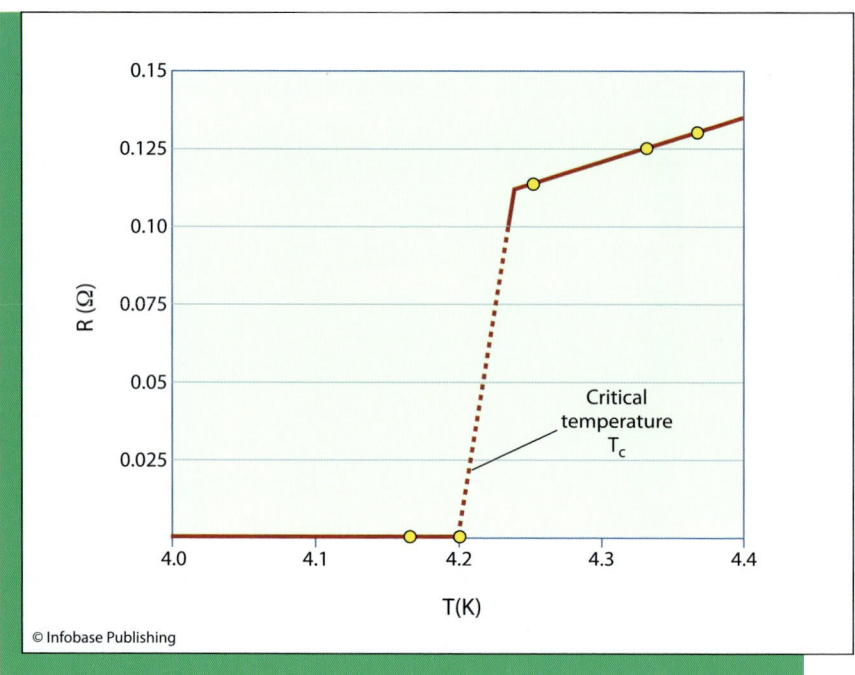

© Infobase Publishing

At the critical temperature, mercury's resistance suddenly drops to zero.

temperature, called the *critical temperature* and often denoted T_c, some substances suddenly lose all resistance and become superconductors. The first substance discovered to have superconducting properties at low temperature was mercury. Mercury's critical temperature is about -452°F (-268.9°C)—4.2 K. The figure above shows a graph of resistance versus temperature for mercury.

The change from ordinary conductor to superconductor is abrupt, and scientists consider it to be a phase transition. Above the critical temperature, the substance is an ordinary conductor, but changes phase into a superconductor at the critical temperature.

With no resistance, superconductors can maintain a current indefinitely, even without the application of voltage. Once an operator initiates a current in a superconductor, the current will last for a long time. These perfect conductors eliminate the waste associated with resistance, which reduces current and transforms electrical energy into heat, which must be dissipated. (Computers, for example, require fans in order to cool electronic components, preventing the heat generated by the electrical circuits from damaging sensitive components.) Remarkable efficiency makes superconductors excellent for many applications, although some materials cannot carry very much current without losing their superconductivity.

Many materials exhibit superconductivity. For example, elements including aluminum, lithium, tin, zinc, tungsten, and lead are superconductors, although elements such as these generally have low critical temperatures. Lead has one of the highest, about 7.2 degrees above absolute zero—-445°F (-265°C). But not all conductors show superconductivity. Copper, silver, and gold are among the best conductors but are not superconductors. The reason for this will be discussed in the following section.

Superconducting cable
(Fermilab Visual Media Services)

In addition to the absence of electrical resistance, superconductors have peculiar magnetic properties. Weak magnetic fields do not penetrate a superconductor; instead, surface currents are set up that negate any internal magnetic field. If a researcher applies a weak or moderate magnetic field in a material and then cools that material below the critical temperature, the internal magnetic field disappears! This phenomenon is known as the Meissner effect, discovered by the German physicist Walther Meissner (1882–1974) in 1933. A strong external magnetic field, however, disrupts superconductivity, and the material becomes an ordinary conductor.

Two types of superconductors have been found, and scientists have given them the not very descriptive names Type I and Type II. Type I superconductors are mainly metals with the properties described above. Type II superconductors are a little more complicated and have a temperature range in which they are in a mixed state—zero resistance but a slight internal magnetic field. Many Type II superconductors are metallic compounds that tend to have higher critical temperatures. An advantage of Type II superconductors is that they can usually carry a lot of current, so they are more often used in industrial applications.

BCS THEORY—A PARTIAL EXPLANATION

In 1957, the American physicists John Bardeen (1908–91), Leon N. Cooper (1930–), and Robert Schrieffer (1931–) proposed a theory to explain superconductivity. This theory has become known as the BCS theory. The name of the theory comes from the initial letter of the last names of the three researchers. BCS theory was the first successful theory of superconductivity, although it does not explain all superconductors.

In an ordinary conductor such as copper, which is often used in circuits such as those that carry electrical power into houses, mobile electrons carry the current. The application of a voltage provides a force that moves the electrons along the conductor. The flow of electrons constitutes a current, the size of which depends on the strength of the voltage. But conductors are not hollow tubes in which electrons flow; the materials are composed of atoms and molecules, in many cases bonded to form a lattice or geometrical arrangement. As electrons move, they bump into the lattice, which impedes their flow. The collisions generate heat, robbing the current of some of its energy. This is the basis of resistance.

Metals have a lot of mobile electrons, which is why these elements are effective conductors. But even a metal such as a copper wire has a certain amount of resistance. Resistance limits current in many materials by a simple formula known as Ohm's law, named after its discoverer, the German physicist Georg Simon Ohm (1787–1854). (The unit of resistance, the ohm, is also named in his honor.) Ohm's law states that current (I) equals voltage (V) divided by resistance (R). In mathematical terms, the equation is $I = V/R$. This "law" is merely a rough approximation of the electrical characteristics of materials and does not always apply.

Physicists realized early on that a unique reasoning was needed to explain superconductors. Even at low temperatures, a material's lattice does not disappear. The absence of resistance in superconductors cannot be understood in terms of the simple motion of electrons.

BCS theory invokes a kind of collaboration between electrons. This relationship is a partnership known as a *Cooper pair*. Consider, for ex-

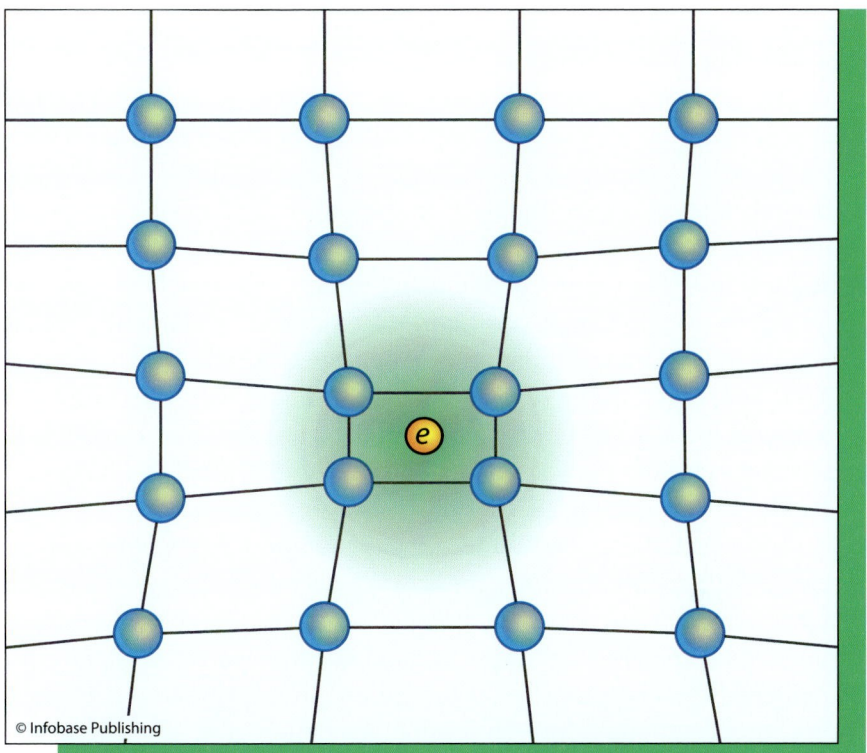

© Infobase Publishing

An electron, represented by the small circle, and its electric field, represented by shading, interact with the lattice, slightly distorting the structure.

ample, an electron in a lattice. The electron's charge is negative, which attracts positive charges and repels other negative ones. A partnership between electrons might seem strange since like charges repel, but the lattice mediates the interaction. An electron attracts positively charged atoms and molecules of the lattice, which creep toward the electron, slightly deforming the lattice structure, as illustrated in the figure above. The deformation concentrates positive charges in the area, creating a surrounding region of positive charge. This region attracts a passing electron. As a result, the electron pair teams up. The deformations are associated with vibrations in the lattice called *phonons,* which form as the electrons navigate through the material. Phonons are like waves or ripples that propagate through the material—the atoms and molecules jostle their neighbors, which in turn jostle other atoms and molecules, propagating the disturbance.

To describe a Cooper pair with mathematical precision, physicists employ quantum mechanics. Cooper pairs form between electrons having equal but opposite values of momentum and a quantum mechanical property known as spin. The electrons become linked into the same state and behave as a single system. Linked in this way, a single quantum mechanical function represents the entire system, which acts as one. The motion of one electron is counterbalanced by that of the other, which cancels all disruptions in their flow—no heat-generating resistance can arise. This effect seems strange but is in keeping with other unusual quantum mechanics concepts, such as the representation of electrons as waves as well as particles. BCS theory accurately predicts energy and magnetic measurements based on these advanced concepts.

Some materials are not suited for superconductivity. Copper, silver, and gold are metals that have some of the lowest resistances of any elements, which means they are excellent conductors. But the lattice structure of these metals is tightly packed and does not lend itself to the kind of interactions necessary for the formation of Cooper pairs. As a result, these metals, although superb conductors, do not generally display superconductivity even at temperatures quite close to absolute zero.

BCS theory neatly accounts for the properties of superconductors having low critical temperatures—the low-temperature superconductors—especially Type I superconductors. But the theory has run into difficulty with more recently discovered superconductors, as described in the section "Beyond BCS Theory—High-Temperature Superconductors."

MEASURING THE BRAIN'S MAGNETIC FIELD

Superconductors are important subjects from which physicists are learning more about the nature of matter and quantum mechanics. But superconductors also enjoy widespread use in technological applications.

One of the most important superconductor applications in physics research involves particle accelerators, the subject of chapter 2. The gigantic magnets needed to steer and accelerate beams of particles come from currents flowing in superconductors—any current, whether in a conductor or superconductor, generates an external magnetic field that affects objects outside of the conductor or superconductor. Magnetic fields generated from superconductors are efficient and controllable and are essential components in today's particle accelerators, despite the expense of keeping the superconductors at extremely low temperatures.

Many other important applications entail the detection of magnetic fields. Magnetism is ubiquitous in electrical circuits as well as in the environment—Earth has an associated magnetic field, generated deep in its core—and measurement of magnetic field strength is often critical. An instrument known as a superconducting quantum interference device (SQUID) can detect even tiny magnetic fields. SQUIDs employ a quantum mechanical phenomenon known as the Josephson effect, named after the physicist Brian Josephson (1940–   ), who predicted it in 1962. The Josephson effect involves the flow or "tunneling" of charges across a thin barrier between two superconductors. Electrical properties of a circuit containing these barriers, or junctions (known as Josephson junctions), are highly sensitive to magnetic fields and indicate the field strength. Liquid helium keeps the superconductor below its critical temperature.

The sensitivity of SQUIDs permits researchers to measure magnetic fields previously beyond detection. One of the most interesting applications is the measurement of magnetic fields arising from a person's head. These fields are due to the activity of brain cells called *neurons,* most of which generate a small current in the process of signaling other neurons. Charges called ions that are floating in the aqueous solution inside and outside of the cell carry this current. The signaling or communication among networks of neurons underlies all mental activity—sensations, thoughts, and muscle activations. Even though the current is minuscule, it generates a magnetic field, as do all currents. The technique of recording these magnetic fields is called magnetoencephalography.

SQUIDs are able to detect these weak fields but will also record any other magnetic field that happens to be in the vicinity. Most of the other magnetic fields are much stronger than the brain's magnetic field; Earth's magnetic field, for instance, is generally millions of times stronger. Researchers who are recording the magnetic fields from a person must shield their laboratory to reduce unwanted fields from drowning out the fields they want to measure.

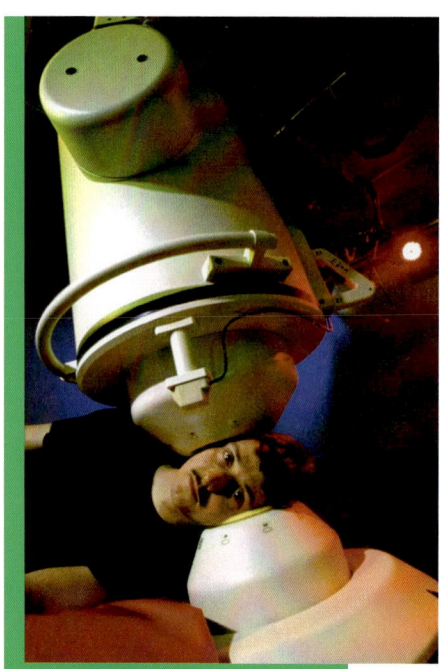

Magnetoencephalography on a human subject *(Dr. Jurgen Scriba/Photo Researchers, Inc.)*

Magnetoencephalography gives scientists and physicians an opportunity to study the electrical activity of the brain without having to perform risky surgery to open the skull. Such noninvasive procedures allow scientists to record the activity of subjects while they engage in meaningful mental activity, such as memorizing a list or performing calculations. Physicians use magnetoencephalography to examine patients for disorders such as epilepsy, in which abnormal activity in the brain causes seizures.

Superconductor applications, while useful, are presently limited. Since critical temperatures are so low, equipment that makes use of superconductors must include a powerful cooling mechanism to maintain the frigid conditions under which superconductors operate. These cooling systems are expensive and usually require special handling and maintenance. If superconductors could function at higher temperatures, they would enjoy vastly more applications and have a much greater impact on society than at present.

BEYOND BCS THEORY—HIGH-TEMPERATURE SUPERCONDUCTORS

The theoretical concepts underlying BCS theory suggested that superconductivity would exist only at extremely low temperatures. Above about -405°F (-243°C)—30 K—BCS theory would be unlikely to hold. To someone who believed that BCS theory was the whole story, this meant that no superconductors existed with a critical temperature above this value.

However, in 1986, Johannes G. Bednorz (1950–) and Karl A. Müller (1927–), researchers employed at IBM Zurich Research Laboratory in Switzerland, discovered a superconductor with a critical temperature slightly higher than the assumed limit. The material was a compound containing barium (Ba), lanthanum (La), copper (Cu), and oxygen (O). Bednorz and Müller published their paper, "Possible High T_c Superconductivity in the Ba-La-Cu-O System," to a skeptical scientific community in a 1986 issue of *Zeitschrift für Physik B Condensed Matter*. Note the qualifier "possible" in their title. This result was so surprising that the discoverers could hardly believe it! But other laboratories soon confirmed the finding.

The falsification of the theoretical limit for critical temperatures was not the only surprising aspect of the 1986 discovery. Superconductors that had been found earlier were generally metals or alloys, which behave as excellent conductors at ordinary temperatures. This newly discovered material belonged to a large class of materials called ceramics, which consist of minerals commonly fired or baked to produce useful products such as pottery and brick. Most ceramics are not effective conductors. While some ceramic materials are semiconductors, many are often used as insulators to block electric currents. Few people expected to find a ceramic superconductor at any temperature.

Bednorz and Müller's finding touched off a wave of interest in the new superconductors. Researchers tested many different ceramic materials, searching for other superconductors. In 1987, Maw-Kuen Wu at the University of Alabama at Huntsville, Paul Chu (1941–) and Pei-Herng Hor at the University of Houston, and their colleagues found a superconductor with a critical temperature of -292°F (-180°C)—93 K. The sidebar on page 102 discusses the superconducting ceramic materials.

The 1987 discovery marked a milestone. This superconductor was the first in which the critical temperature exceeded the boiling point of liquid nitrogen, which is -321°F (-196°C), or 77 K, at atmospheric pressure. Liquid nitrogen is often used as a coolant, commonly found in cooling systems such as those involved in transporting frozen food products. Because the superconductor's critical temperature was high enough, it could be cooled with liquid nitrogen.

Researchers scrambled to find superconductors with increasingly higher critical temperature. The record as of June 2009 is a compound

Superconducting Ceramic Materials

The materials used in the experiments of Bednorz and Müller, and Chu, Wu, and their colleagues, belong to a category called cuprates, which are compounds containing copper oxide. These materials are also members of an abundant family of minerals known as perovskites. Researchers often refer to these superconductors as cuprate superconductors or cuprate-perovskite superconductors. Bednorz and Müller worked with a compound of the elements barium, lanthanum, copper, and oxygen in 1986, while the compound of Chu, Wu, and their colleagues substituted the element yttrium for lanthanum. Researchers commonly call the compound yttrium barium copper oxide by its initials, YBCO.

Chemical formulas for these compounds demonstrate their complexity. The first YBCO, for instance, had the formula $YBa_2Cu_3O_7$. Other materials are equally complex. Another compound, whose critical temperature is -234°F (-148°C)—125 K—has the formula $Tl_2Ba_2Ca_2Cu_3O_{10}$. (Tl is the symbol for thallium, and Ca stands for calcium.) These compounds have complicated crystal structures and consist of multiple layers. Crystals of this kind are not easy to produce.

Superconducting ceramic materials have certain drawbacks when compared to metallic superconductors. As anyone who has ever handled pottery knows, a common char-

containing mercury, thallium, barium, calcium, copper, and oxygen that has a $T_c = -211°F$ (-135°C), or 138 K. In some cases, scientists can achieve higher critical temperatures by raising the pressure.

Ceramic superconductors fall into the Type II class. As in all superconductors, the electrons carrying the current manage to navigate their way through the material without losing energy from colliding or bumping into the constituent atoms and molecules. BCS theory, which neatly accounts for low-temperature metallic superconductors, ascribes acteristic of ceramics is brittleness—ceramic materials are hard and inflexible and shatter when dropped. They are also not easily fashioned into various shapes as needed, whereas metals and alloys are extremely shapeable. Their one advantage is the relatively high temperatures at which they can function, which makes superconducting ceramic materials at least a stepping-stone toward more widespread applications.

Magnetic levitation—magnetic properties of this YBCO ceramic superconductor levitate a magnet *(David Parker/IMI/University of Birmingham/Photo Researchers, Inc.)*

the absence of resistance to the formation of Cooper pairs. These pairs of electrons form a system that traverses the lattice with the mediation of vibrations called phonons. Momentum and spin of electron pairs cancel, creating a coherent quantum mechanical state.

But BCS theory did not seem to apply to the higher temperatures and the complicated structures of the new superconductors. Physicists believed that electron pairs formed in the cuprate superconductors, and these pairs team up to glide effortlessly through the material, but the resulting quantum mechanical state was slightly different. William A. Little, a researcher at Stanford University, wrote a 1988 *Science* paper, "Experimental Constraints on Theories of High-Transition Temperature Superconductors," in which he discussed the problem. Many of the energy and magnetic properties of the new superconductors conform to BCS theory. "The one difference from a conventional BCS superconductor appears to be the mode of coupling. Evidence suggests that some charged excitation, with an energy several times that of phonons, provides this coupling. This remains to be identified."

Finding a correct theory, or modifying BCS so that it applies to high-temperature superconductors, is important for several reasons. Scientists want to understand how the world works, and theories help them to extend their observations to the general case—a theory elegantly summarizes what scientists have learned.

Theories also generate predictions. Being able to make accurate predictions is crucial when researchers are searching for rare events or materials, for otherwise researchers must rely on trial-and-error observations. Consider the millions of substances known to exist; researchers would like to know which, if any, are the most promising high-temperature superconductors. Without theoretical guidance, researchers do not know where to look, and a random search is time-consuming and unlikely to be successful.

CONSTRUCTING A NEW THEORY

A theory explaining high-temperature superconductors needs to account for the coupling of the electrons, by which the particles overcome their electrical repulsion. In BCS theory, phonons mediate the coupling. If phonons do not mediate the coupling in the high-temperature superconductors, physicists must discover what does.

Several alternatives have emerged. Two important properties of electrons—their magnetic field and spin—could potentially provide the necessary interactions. Instead of riding phonons, electrons might be navigating superconductors with the aid of fluctuations in magnetism or spin.

The concept of spin does not refer to rotation. Spin is a quantum mechanical concept, involving the mathematics of quantum physics. Electrons, for instance, are pointlike particles—physicists have not found any structure in electrons—yet they have spin, as well as mass, charge, and magnetic fields, and also behave in certain circumstances as waves.

Magnetic or spin fluctuations could convey electron pairs in a manner similar to that of lattice vibrations, although these fluctuations would constitute a different mechanism. The magnetic or spin interactions of electrons would be the source of the pairing and the fluctuations within the material would be the "wave" on which the electron pairs "surf" through the superconductor unimpeded. Physicists use advanced mathematics to formulate a more precise description of this activity.

Learning the mechanism or mechanisms responsible for the electron coupling is critical. When researchers discover how the high-temperature superconductors work, they will be able to focus their search for new superconductors—those with even higher critical temperatures—on the materials whose properties are the most favorable.

But superconductors have been a source of much surprise in the last few decades, and there may be plenty of surprises left in store. Some researchers are beginning to find that perhaps people were too quick to dismiss phonons in the high-temperature superconductors.

In 2004, Alessandra Lanzara, a researcher at the University of California, Berkeley, and the Lawrence Berkeley National Laboratory, and her colleagues reported an experiment in which they tested for the importance of phonons in high-temperature superconductivity. The researchers used high-quality crystals in which the isotope oxygen-18 was substituted for the more common isotope oxygen-16. This substitution gave the lattice more mass, which affected its vibration. Lanzara and her colleagues performed measurements with advanced spectroscopic equipment and found that the lattice influenced the conduction of the electrons in the superconductor. In other words, there was an important interaction between the electrons and the lattice.

A press release issued by the University of California, Berkeley, on August 16, 2004, quoted Lanzara on the significance of the finding, "The results we found provide the first direct evidence for a significant and unconventional role of phonons in the high temperature superconductivity, meaning that all the reasons that have been used so far to disregard the importance of phonons are not valid anymore."

This experiment did not prove that phonons were the sole mediators of the electron coupling, and the researchers acknowledge that magnetic or spin fluctuations may play a role. Spin interactions, for example, could enhance the coupling, making it tighter. This effect could in turn strengthen the electron-phonon interaction. When combined, the mutually reinforcing effects could yield superconductivity in materials in which either alone would not suffice.

But other researchers have obtained results that seem to muddy the picture. In 2006, Tonica Valla, a physicist at the Brookhaven National Laboratory in New York, presented some new experimental findings at a meeting of the American Physical Society, a scientific organization devoted to physics. Valla and his colleagues have observed both phonons and spin fluctuations in superconductors. To clarify the issue, he and his team began observing materials that are not superconductors, yet are made of the same elements as the high-temperature superconductors.

Consider the first high-temperature superconductor, which was made of lanthanum, barium, copper, and oxygen. Scientists often adjust the components of complex materials by adding or removing small numbers of atoms; this process, called "doping," can significantly alter the properties of the materials and is frequently used, for example, in the manufacture of semiconductors. Researchers can use doping to manipulate the properties of superconductors, including their critical temperature. And with certain ratios of atoms in the high-temperature cuprate crystals, they lose their superconductivity.

Valla and his colleagues decided to compare the material in its superconducting and nonsuperconducting states. The differences would presumably be instrumental in affecting the change between states and would help to resolve which interactions are critical in the process. Valla's team used spectroscopy as well as advanced techniques to analyze the state of the electrons in the crystal. For example, when the researchers train an intense beam of ultraviolet radiation on the material, it emits electrons. By measuring the electrons' velocity and the angles at which they leave, the scientists can calculate what the electrons were doing inside the material.

But Valla and his colleagues found that the properties of the material when it is not a superconductor are not very different than when it is. "The fact that this system, which is not a superconductor, has similar properties to the superconducting system is not helping to solve the mystery," Valla was quoted as saying in a press release, "New Wrinkle in the Mystery of High-Temperature Superconductors," posted on Science*Daily* on March 21, 2006. "We are still at the beginning," he said. But he noted, "It looks like the story is getting more complicated."

More data, including perhaps data from new materials, will add to the story. The story is likely to get even more complicated until the answer finally emerges. But this is often the case at the frontiers of science. In the meantime, the hunt for new materials, although hindered without a comprehensive theory, continues apace.

SEARCHING FOR NEW MATERIALS

A team of Japanese scientists led by Hideo Hosono (1953–) of the Tokyo Institute of Technology announced an important discovery in February 2008—a new family of superconductors. While working with iron compounds, Hosono and his colleagues hit upon the right ingredients, finding that lanthanum oxygen fluorine iron arsenide exhibits superconductivity at -413°F (-247°C), or 26 K.

Given the magnetic properties of superconductors, the existence of iron in a superconductor surprised researchers. Iron is ferromagnetic, meaning that it can remain magnetized when exposed to an external magnetic field—this is how a piece of iron can become a magnet. Magnetic fields tend to reduce or even eliminate superconductivity, yet a substance containing iron arsenide—iron and arsenic—is a superconductor. Iron arsenide superconductors open the possibility for superconducting materials capable of tolerating higher currents and stronger magnetic fields, which would enhance their present applications.

Researchers scrambled to replicate the experiment. In March 2008, X. H. Chen at the University of Science and Technology of China discovered samarium oxygen fluorine iron arsenide becomes a superconductor at -382°F (-230°C), or 43 K, and Zhong-Xian Zhao of the Institute of Physics at the Chinese Academy of Sciences in Beijing found that praseodymium oxygen fluorine iron arsenide is a superconductor at -366°F (-221°C), or 52 K. Although the iron arsenides are not setting temperature records, they have become another important class of superconductors, distinct from the cuprates.

Discovery of this new family of superconductors raises several important issues. Physicists would very much like to know if the mechanism of these superconductors is the same as the cuprates. Research on the new iron arsenide superconductors, as well as continuing studies on the cuprates, should help to uncover the mystery of high-temperature superconductivity.

A collaboration of scientists from Oak Ridge National Laboratory in Tennessee, Ames Laboratory in Iowa, California Institute of Technology, Argonne National Laboratory in Illinois, and Rutherford Appleton Laboratory in the United Kingdom have begun to probe the new superconductors with a sophisticated tool. Their experiments involve studying the superconductor's activity by sending a beam of neutrons into the material. Neutrons are components of the atomic nucleus, and free neutrons interact with nuclei, engaging in reactions as well as other activities. The magnetic properties of neutrons also affect, and are affected by, the spin of the material's components.

Neutrons are difficult particles with which to work. In addition to their strong interactions with nuclei—free neutrons do not stay that way for long—neutrons are electrically neutral, which means they cannot be accelerated with the electromagnetic techniques that work so well with protons and electrons. Researchers often generate neutron beams by accelerating other particles and smashing them into targets; the collisions kick out neutrons, which researchers can focus into a beam by guiding them through narrow openings.

The collaborative team studying the new superconductors is making use of a gigantic complex known as Spallation Neutron Source, located at Oak Ridge National Laboratory. This $1.4 billion facility, completed in 2006, creates intense beams of neutrons by accelerating protons and directing them into a target of mercury. The protons interact with mercury nuclei in a reaction called spallation, in which neutrons are ejected. (The verb *spall*, which is not commonly used these days, means to break or make smaller by chipping.) Conduits channel the ejected neutrons, whose speed may be slowed by passing them through water or liquid hydrogen.

Spallation Neutron Source, along with other facilities and instruments, put Oak Ridge National Laboratory in an excellent position to study advanced materials such as superconductors. The sidebar on page 109 provides more information on this important national laboratory.

On October 10, 2008, Ames Laboratory issued a press release describing some of the earliest findings of the team of researchers using Spallation Neutron Source. The team included Robert J. McQueeney of Ames Laboratory and Andrew D. Christianson of Oak Ridge National Laboratory. Studying the iron arsenide superconductors, researchers

Oak Ridge National Laboratory

The U.S. government generally funds national laboratories in order to conduct research that is in the national interest and requires too much money or is too dangerous for private institutions to undertake. One such program was the Manhattan Project, the 1940s project that developed the atomic bomb to end World War II. Among the facilities participating in this top-secret project was one built in 1943 on an isolated patch of land in the mountains and ridges of eastern Tennessee. Originally called Clinton Laboratories—named after the closest town—this facility housed 75,000 residents working feverishly on supplying radioactive materials for the development of the atomic bomb.

After World War II ended in 1945, so did the laboratory's nuclear weapons mission. Researchers shifted to other research, but the laboratory maintained its expertise in particle physics and isotopes. In 1948, the laboratory received its present name, and researchers began working on several programs, including the physics and technology of electricity-generating nuclear reactors and the production of radioactive isotopes for the treatment of cancer.

Today the staff numbers about 4,200. Oak Ridge National Laboratory scientists enjoy one of the world's largest neutron facilities, including the Spallation Neutron Source and High Flux Isotope Reactor. As part of the Department of Energy national laboratory system, Oak Ridge National Laboratory has many programs concerning the development of alternative energy sources and the improvement of existing systems, including the potential use of superconductors to reduce losses in electric power transmission. The laboratory also has projects involving nanotechnology—the design and use of molecular-sized tools and technologies—as well as the design of sensors to detect illicit nuclear materials.

analyzed neutrons that passed through and interacted with nuclei in the superconductor. The velocity and angle of the neutrons offered clues about the position and state of these nuclei, which tells scientists something about the superconducting properties of the material.

In the press release, McQueeney said, "The preliminary results are amazing." The instruments worked well, providing a lot of data in a short period of time. With this data, the researchers examined what the motion and state of the nuclei could tell them about the unimpeded passage of the electron pairs. "Our measurements did not support the conventional electron-phonon mediated superconductivity," McQueeney noted. The vibrations in the lattice did not seem to be involved in the electron coupling. Researchers came to this conclusion when they studied the energy of the neutrons scattered from the superconducting material and found little association with the movement of electrons.

Research on the newly discovered iron arsenides is only in the initial stages. Future experiments may support or contradict the early findings as researchers bring other techniques and instruments to bear on the problem. Other important questions to answer involve the possibility that these new materials, or materials derived from them, will eventually produce superconductors with even higher critical temperatures than has already been achieved with cuprates.

CONCLUSION

The new family of superconductors gives researchers an opportunity to study the phenomenon in yet another material. Such opportunities often lead to advances at the frontiers of science, since researchers can compare and contrast properties, winnowing out the irrelevant ones and sharpening the focus on the critical ones.

But no one presently knows if any of these materials, or a material yet to be discovered, can exhibit superconductivity at warm temperatures—room temperature, say—that would vastly enhance their applicability. A superconductor that could operate in the environment with no expensive cooling requirements would have tremendous benefits.

Consider power transmission, for example. Utility companies generate electricity at power plants and then distribute the power to homes and businesses located at various distances from the plant. The current travels in overhead or underground transmission lines. Since the wires that compose these transmission lines are not perfect conductors, they have a certain amount of resistance. This resistance wastes part of the energy by turning some of it into heat. The use of very high voltage minimizes these losses but makes the power lines dangerous and also requires transformers to reduce the voltage to a level appropriate for consumers. Losses incurred during transmission depend on the distance, varying from only a few percent for short distances to more than a quarter in rural distribution.

Superconducting transmission lines would eliminate this enormous waste. Although the presently available superconductors would do the job, the expense of the necessary cooling systems more than offsets the benefits. To reduce this waste, superconductors with higher critical temperatures are needed.

Some researchers are turning to computers to guide them in the search for new superconductors. An international team led by Guoying Gao and Guangtian Zou of Jilin University in China and Artem R. Oganov at the Eidgenössische Technische Hochschule (ETH) Zürich (Swiss Federal Institute of Technology) announced the results of a computational calculation in 2008. The researchers used a sophisticated algorithm in which the theoretical behavior of particles composing a specific material—germanium hydride (GeH_4)—was calculated based on advanced physics. As described in a press release issued by ETH, calculations indicated that germanium hydride would be a superconductor with a critical temperature at -344°F (-209°C), or 64 K.

Having a new superconductor to study is important, and this compound is much easier to manufacture. But in terms of applications, a superconductor at this temperature is not much of an advance. It does not even reach the boiling point of liquid nitrogen, so this common cooling mechanism could not be employed. The researchers believe it may be possible to raise the critical temperature a few degrees by doping the material with tin or silicon, but even so, the calculations dictate that germanium hydride must be under extremely high pressure in order to become a superconductor. Pressure required for the transition is about 2 million times that of the atmosphere at sea level.

Computer algorithms to predict the properties of materials are helpful in superconductor research. But the programs are time-consuming to run and, as is the case with experimental searchers for new superconductors, proceed one material at a time. A comprehensive theory would be vastly superior because it would provide specific guidance.

Physicists do not yet know if a room-temperature superconductor is possible, and if so, what kind of material would be needed. Lacking an adequate theory of high-temperature superconductivity, researchers who seek a room-temperature superconductor must continue to develop experiments and computer algorithms in the hope of a lucky breakthrough. In the meantime, physicists are studying these materials to gain a theoretical understanding that will provide intellectual satisfaction as well as much-needed guidance.

CHRONOLOGY

1827	The German physicist Georg Ohm observes that the voltage across a circuit element is proportional to the product of the current and resistance, a relationship known as Ohm's law. This "law" holds true for many materials at a wide range of temperatures.
1908	The Dutch physicist Heike Kamerlingh Onnes develops the technology to reach temperatures cold enough to liquefy helium.
1911	Onnes discovers superconductivity—mercury loses its resistance at -452°F (-268.9°C)—4.2 K.
1933	The German physicist Walther Meissner discovers that materials lose their interior magnetic field when they become a superconductor.
1957	The American physicists John Bardeen, Leon N. Cooper, and Robert Schrieffer propose the BCS theory to explain superconductivity.
1962	The British physicist Brian Josephson (1940–) predicts an important effect, later called the Josephson effect, which involves the tunneling of charges through a barrier.

Researchers at Westinghouse develop a niobium-titanium superconducting wire, important for many superconductor applications.

1964 Ford Scientific Research Laboratory scientists develop the first superconducting quantum interference device (SQUID).

1986 Johannes G. Bednorz and Karl A. Müller, researchers at IBM Zurich Research Laboratory in Switzerland, discover a high-temperature superconductor.

1987 Paul Chu at the University of Houston and his colleagues find a superconductor with a critical temperature of -292°F (-180°C)—93 K—higher than the boiling point of liquid nitrogen, a common coolant.

Fermi National Accelerator Laboratory's Tevatron is the first particle accelerator to employ special superconducting magnets.

2004 Alessandra Lanzara, a researcher at the University of California, Berkeley, and the Lawrence Berkeley National Laboratory, and her colleagues find evidence of phonon involvement in high-temperature superconductors.

2008 A team of scientists led by Hideo Hosono of the Tokyo Institute of Technology announce the discovery of an iron arsenide superconductor.

FURTHER RESOURCES
Print and Internet

Ames Laboratory. "New Instrument Puts New Spin on Superconductors." News release (10/10/08). Available online. URL: http://www.

ameslab.gov/final/News/2008rel/Iron-arsenic_superconductors. html. Accessed June 22, 2009. The Spallation Neutron Source shows promise in uncovering the mechanisms of high-temperature superconductors.

Argonne National Laboratory. "Superconductivity Center at Argonne National Laboratory." Available online. URL: http://superconductivity. et.anl.gov/. Accessed June 22, 2009. Argonne National Laboratory is an active player in the field of superconductors. This Web resource offers information on the laboratory's techniques, facilities, publications, and patents.

Bednorz, J. G., and K. A. Müller. "Possible High T_c Superconductivity in the Ba-La-Cu-O System." *Zeitschrift für Physik B Condensed Matter* 64 (1986): 189–193. The report of the initial discovery of high-temperature superconductors.

Eck, Joe. "Superconductors." Available online. URL: http://www.super conductors.org/. Accessed June 22, 2009. Plenty of information on all aspects of superconductors are available at this well-managed Web resource. Topics include Type I and Type II superconductors, the history of the subject, atypical superconductors, terminology, applications, and news about the latest research.

ETH Zürich. "Crystallographers Use Computers to Find a New Superconductor." News release (12/1/08). Available online. URL: http:// www.ethlife.ethz.ch/archive_articles/081201_oganov_paper_nsn/ index_EN. Accessed June 22, 2009. This research involves the use of a computational algorithm to predict the superconducting properties of germanium hydride.

Ginzburg, V. L., and E. A. Andryushin. *Superconductivity.* Singapore: World Scientific Publishing, 2004. The authors are two researchers who provide a nontechnical overview of superconductors, including their history, physics, and applications.

Kirkland, Kyle. *Electricity and Magnetism.* New York: Facts On File, 2007. Part of the Physics in Our World set, this volume discusses the physics of electricity and magnetism, including important applications such as superconductivity, on a level accessible to high school students.

Lawrence Berkeley National Laboratory. "Getting Kinky with High-Temperature Superconductors." Available online. URL: http://www.

lbl.gov/Science-Articles/Archive/ALS-kinky-conductors.html. Accessed June 22, 2009. This article, posted on January 14, 2002, describes some important experiments in high-temperature superconductivity and phonons.

Little, W. A. "Experimental Constraints on Theories of High-Transition Temperature Superconductors." *Science* 242 (12/9/88): 1,390–1,395. In this early article on the theory of the cuprate superconductors, Little describes the experimental measurements that mark similarities and differences with BCS theory.

Oak Ridge National Laboratory. "High-Temperature Superconductivity for Electric Systems." Available online. URL: http://www.ornl.gov/sci/htsc/. Accessed June 22, 2009. This Web resource provides news and information on Oak Ridge National Laboratory's research on high-temperature superconductors.

Science*Daily*. "New Wrinkle in the Mystery of High-Temperature Superconductors." News release (3/21/06). Available online. URL: http://www.sciencedaily.com/releases/2006/03/060317114140.htm. Accessed June 22, 2009. The findings of Tonica Valla, a physicist at Brookhaven National Laboratory, and his colleagues show that superconducting and nonsuperconducting material can have similar properties.

Trigg, George L. *Landmark Experiments in Twentieth Century Physics.* Mineola, N.Y.: Dover Publications, 1995. The experiments discussed in this historical account include the discovery of X-rays, superconductivity, transistors, laser, quantum tunneling, neutrinos, and many others.

University of California, Berkeley. "Vibrations in Crystal Lattice Play Big Role in High Temperature Superconductors." News release (8/16/04). Available online. URL: http://www.berkeley.edu/news/media/releases/2004/08/16_Lanzara.shtml. Accessed June 22, 2009. Alessandra Lanzara and her colleagues find evidence that phonons are involved in the mechanism of high-temperature superconductors.

5

CHAOS THEORY AND THE BUTTERFLY EFFECT

Mathematics plays an essential role in physics. Measurements are quantified, as they are in most sciences, but physicists have been particularly successful in finding equations and formulas that relate the various measurements. The value of a variable, such as energy, E, can be related to an object's mass, m, in the equation, $E = mc^2$, where c is the speed of light in a vacuum. This equation featured prominently in chapters 1, 2, and 3, and has served physicists well since Albert Einstein discovered it in 1905. But most of the mathematics in physics is much more complicated.

One of the most recently developed mathematical concepts in science is chaos theory. Chaos in "chaos theory" does not refer to the conventional definition of the term, which means disorder or a state of confusion. What chaos theory involves is a mathematical description of the behavior and evolution of a *dynamical system*—a system that is dynamic, meaning the components change over time. In particular, chaos theory deals with systems in which small changes or perturbations can have drastic effects. This is true of weather, which provides an example of perhaps the best-known effect of chaos—the butterfly effect. The butterfly effect refers to the notion that the tiny perturbation caused by the flapping of a butterfly's wings in South America, for example, could lead to tremendous consequences in the atmosphere, perhaps instigating a tornado in the United States. This degree of sensitivity may seem unrealistic, but systems exhibiting chaos in the mathematical sense are highly susceptible to changing conditions.

Applications of chaos theory exist in almost every science, but physics is particularly rich. Physicists study the motion, properties, and evolution of dynamical systems of all sizes, from fusing nuclei to orbiting planets. Being trained in mathematics also helps physicists to use and apply advanced mathematical concepts such as chaos theory. Many physicists along with mathematically minded colleagues also study chemistry and biology.

Although the term *chaos* suggests disorder, chaos theory and its applications are just the opposite—in mathematics and science, chaos concerns order rather than disorder. Some systems appear random and disorganized, yet they are not quite as disordered as they seem. Since the 1960s, scientists have been studying chaotic systems to learn how to spot order in what seems to be disordered behavior. This chapter describes the basic principles of chaos theory and explores how scientists are using chaos theory to get a better handle on a variety of complicated systems.

INTRODUCTION

The French mathematician Henri Poincaré (1854–1912) discovered the mathematical concepts of chaos in 1890 while he was working on a problem in astrophysics known as the three-body problem. This problem concerns finding the trajectories of three objects that are interacting according to laws discovered by Sir Isaac Newton: the laws of motion and the universal law of gravitation, which says that the gravitational attraction between two bodies is proportional to the product of their masses and inversely proportional to the distance between them. For example, some star systems contain three stars, and Newton's laws govern their orbits.

Solving these problems requires calculus, and the problem gets complicated when there are more than two interacting objects. The two-body problem is not difficult to solve, but the calculation involving three or more bodies generally has no solution that can be easily formulated. Researchers can find approximate solutions to these problems—and scientists of the modern era can program computers to do so quickly—and Poincaré found certain trajectories or orbits that had interesting features. Some of these paths were bounded, or, in other words, limited to a certain region of space, yet the paths failed to be periodic—they did not return to the starting point. Objects moving in such paths would appear to be extremely disordered, unlike a periodic orbit in which the object repeatedly followed the same route. Poincaré later conjectured that predicting the path of these orbits would be nearly impossible because the system was so sensitive to any perturbation.

But chaos theory did not receive widespread attention until the 1960s. In 1961, Edward Lorenz (1917–2008), a meteorologist at the Massachusetts Institute of Technology (MIT), was using a computer to find the solutions to a set of three equations describing the flow of fluid. The computer performed an integration—a method of solving the equations—step-by-step, so that the computer's output showed the progression over time of the motion, from start to finish. Lorenz set up the computer so that he could input the starting values of the relevant variables—the *initial conditions*—and then let the program run through the calculation, computing the value of these variables at each point as the system evolved over time. Kerry Emanuel, an MIT researcher, described in a 2008 *Science* article what Lorenz found: "Wanting to carry the integration further in time, he [Lorenz] re-started a calculation at about the midpoint of his first run, using the numerical output as his starting state. Escaping the racket of the machine [computers at the time could be quite noisy], he stepped out for a cup of coffee, but on returning found that the solution had diverged greatly from the first run. At first suspecting a machine malfunction, he quickly realized that he had stumbled on a proof of Poincaré's conjecture: On reentering the data, he had merely rounded the output to three significant figures."

Lorenz had discovered the system's sensitivity to initial conditions. By rounding the numbers, he had changed their values by a tiny amount. Yet this change was enough to cause the variables to evolve in a significantly different manner. Small perturbations resulted in a drastic change in the system's subsequent behavior; instead of passing through one set of states, it went through an entirely different set.

In 1963, Lorenz published his observations. But as Emanuel noted, "The article went almost unnoticed outside the atmospheric sciences for nearly a decade." Yet it was supremely important: "This and subsequent work on the mathematical properties of chaotic systems has been called the third scientific revolution of the 20th century." (The other two are quantum mechanics and Einstein's relativity theory.)

Lorenz presented a talk in 1972 titled, "Predictability: Does the Flap of a Butterfly's Wings in Brazil Set Off a Tornado in Texas?" The idea was that even an insignificant event can produce drastic changes in a complicated system such as Earth's weather. If the butterfly had

Weather systems such as Hurricane Felix, photographed from the *International Space Station* on September 3, 2007, arise from complicated interactions in the ocean and atmosphere. *[Science Source/Photo Researchers]*

not flapped its wings, the system would have evolved differently, and perhaps the tornado would not have formed in Texas. Such extreme sensitivity to perturbations is known as the butterfly effect, and it is the essence of chaos theory. MIT released an obituary of Edward Lorenz in 2008, calling him the "father of chaos theory and butterfly effect."

Due to the effects of chaos, complicated systems such as the weather are not easy to predict very far in advance. Predicting tomorrow's weather is possible (although such predictions are not always accurate), but predicting the weather a week or two in advance is highly uncertain. The reason for this is that small, unforeseen events alter the system's behavior so that it diverges from predicted behavior. Another problem is that nobody can measure the initial conditions of a complicated system with perfect precision; any error, no matter how small, in the measurement of the initial conditions means that the predictions based on these conditions will differ from the actual behavior.

Some scientists called chaos theory a major revolution because it seemed to mark an abrupt change in scientific philosophy, as did quantum mechanics and Einstein's relativity theory. Physicists after Newton had conceived of the universe as a predictable system that behaved with clockwork precision and order. But Einstein's relativity theory blasted notions of absolute space and time, and quantum mechanics introduced the concept of probability—the behavior of a system in quantum mechanics is not determined but can only be described by probabilities or tendencies to move into one state or another. Chaos theory is yet another serious blow to simplicity and predictability.

But unlike quantum mechanics, chaotic systems are deterministic, not probabilistic. The future course of a deterministic system is completely defined by the equations that govern its behavior and the value of the initial conditions. In chaos theory, unpredictability is the result of drastic effects arising from small perturbations in the conditions. This unpredictability is not the same as the inherent probabilistic nature of quantum mechanics. The relationship between chaos and quantum mechanics is discussed in a following section "Chaos and Quantum Mechanics."

Not all systems exhibit chaos. A precise mathematical definition of chaos is not easy to understand without mathematical training and will not be presented here. But chaos can be characterized by relatively simple concepts—chaotic systems are deterministic and are extremely sensitive to perturbations.

Systems that show chaotic behavior are always nonlinear, which means that the equation or set of equations describing their behavior are nonlinear. In a linear equation, no variables are raised to a power other than 1, and variables are not multiplied together; for example, the equations $y = x + 2$ and $y = 4x$ are linear equations, but $y = x^2$ and $xy = 4$ are not. A graph of the solution to a linear equation such as $y = x + 2$ is a straight line in an x - y coordinate system. A nonlinear equation is simply any equation that is not linear. Linear systems, and a few nonlinear systems, are generally well behaved, meaning the equations are solvable and the system is easily predictable. But nonlinearities represent complicated interactions that can lead to chaotic behavior.

Note that the system need not involve a lot of variables, which is one of the most remarkable features of chaos theory. Many people are not surprised that large systems such as weather systems, which depend on a huge number of variables, exhibit chaos. But systems with only a few variables can be chaotic as well.

FINDING ORDER IN NONLINEAR DYNAMICS

Chaotic systems appear complicated and disordered. When the system involves the motion of particles, then the trajectories of these particles appear to wander aimlessly. Other systems are composed of variables representing all kinds of states or measurements, such as the electrical activity of a brain cell or the growth of certain populations of living organisms. In any case, the value of the variables evolves and changes in what appears to be a disorderly manner.

Scientists describe and study a dynamical system by constructing a mathematical abstraction known as a *phase space.* The American mathematician and scientist Josiah Willard Gibbs (1838–1903) developed the idea of representing systems in this way in 1901. A phase space,

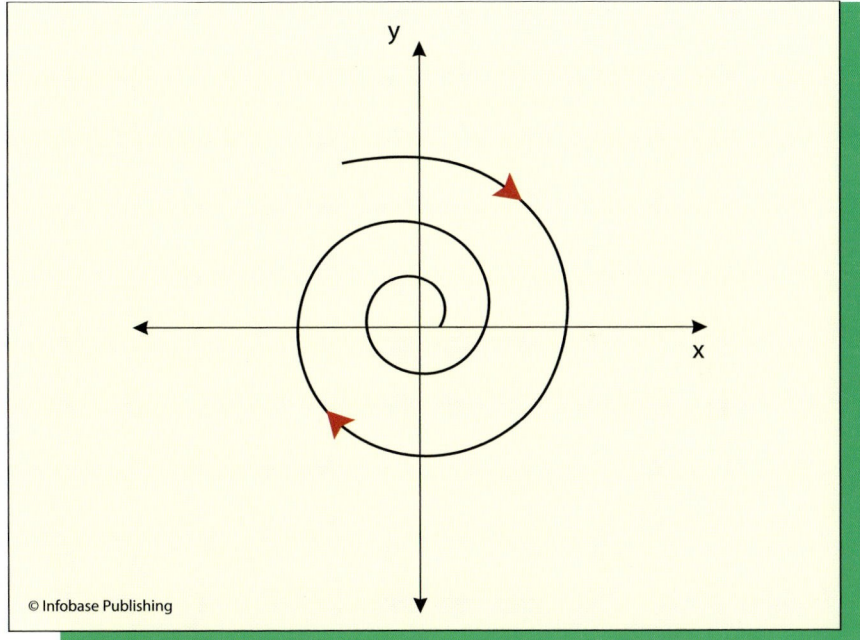

© Infobase Publishing

This phase space diagram represents a system with two variables, *x* and *y*. At each point in time, the values of the two variables are plotted as a point in space. These points form a curve, or trajectory, as the system evolves. In this system, the value of *x* and *y* both approach 0, so the trajectory spirals toward the origin.

which is sometimes called a state space, contains all the possible states of the system. The space is similar to a coordinate system in which each axis represents one of the variables. For example, the figure on page 121 illustrates a phase space of two variables. The axes are perpendicular. A system described by four variables would have a phase space with four axes, although no one can visualize more than three perpendicular axes. The number of variables in a system is often referred to as the system's dimension. High-dimensional systems cannot be visualized, but the mathematical principles of their construction and use still hold.

Phase spaces track the evolution of a system. Each point in time is represented by the value of the system's variables at that time, as marked in the phase space. For example, if a system has two variables, x and y, and $x = 2.5$ and $y = 3.8$ initially, then the (x, y) point of $(2.5, 3.8)$ is marked in phase space. The values of the variables at each instant in time are subsequently marked, forming a trajectory or path in phase space, as shown in the figure. Note that this trajectory is not a depiction of actual movement in real or physical space. A trajectory in phase space shows the values of the variables as the system evolves or changes in time, it does not necessarily show the movement of actual particles. The variables can be any measurement, such as height, weight, temperature, position, momentum, and so on.

A phase space can show each possible progression or evolution that a system may follow. For example, the system may start at point p and evolve to point q, where it settles for a long period of time, perhaps indefinitely. This would be the case for a system that reaches equilibrium and no longer changes. If the system starts at point r, it will take a different path, but it may still settle at the same point q. For example, think of a ball rolling around in a bowl, with the phase space consisting of the ball's position and momentum. The ball could start at any point and be given a certain momentum, but will generally always settle at the bottom (unless the momentum carries it out of the bowl), which will be represented by the point in phase space where the position is at the bottom and the momentum is 0.

Other systems may fail to come to rest at a specific point. Some systems are periodic, changing through a series of states repeatedly. Consider the position and momentum of the pendulum of a clock, for example. Trajectories in phase space of these systems form a closed loop, also known as an orbit. As the system evolves, the variables cycle through this loop.

Sensitivity to initial conditions is easy to illustrate in phase space. Suppose a system starts in point p_1 and evolves in a certain trajectory. If the system is deterministic, it will always evolve in the same trajectory from this point. Now suppose the system starts at a nearby point p_2, in which the values of the variables are almost (but not quite) the same as the first starting point, and an observer plots this trajectory in the same space, without erasing the first trajectory. In some systems the trajectories may converge to the same point, and in other systems the trajectories may parallel one another, following two closely spaced orbits. But in some systems the trajectories quickly diverge, growing widely apart. This divergence indicates sensitivity to initial conditions—even though the systems began at almost the same state, they evolved in vastly different ways. This sensitivity to initial conditions is a hallmark of chaos.

Researchers studying chaos noticed that in some systems, order emerged in the otherwise disorderly phase space trajectories. Chaotic systems fluctuate, often wildly due to their sensitivity, yet the trajectories sometimes collect or gather in certain regions of phase space. A point or region to which a system converges is known as an *attractor.* An attractor can be thought of geometrically as a basin of "attraction" in phase space. When a system begins near an attractor, or evolves close to it, then it will tend to stay there, with the values of the variables remaining the same (if the attractor is a point) or in a small range, as described in the sidebar on page 124.

A system may show chaotic behavior in all or only a portion of its phase space. In any case, the disorderly behavior may have a hidden order and structure—a strange attractor or attractors in which the system tends to get "stuck." Note that sensitivity to perturbations still applies; a trajectory that starts at p_1, for example, will diverge from p_2 even if both are near a strange attractor. The trajectories may stay in the same vicinity, but they will ramble about on their own in the attractor basin, sometimes moving to opposite sides and sometimes briefly moving close together, only to part again.

CHAOS IN THE BRAIN

One of the most complicated systems scientists have discovered is the human brain. Capable of learning complex tasks, formulating scientific theories, inventing new things, and designing cathedrals and computers, the brain processes, stores, and interprets information from a variety of sensory sources. Scientists have been studying the brain for centuries but have yet to arrive at a deep understanding of how the brain gives rise to

Attractors

A simple example of an attractor is a fixed point. A ball rolling around in a bowl, for example, will end up at the bottom, and the point in phase space representing this state is an attractor. For a system exhibiting periodicity, such as a swinging pendulum or a group of chemical reactants that cycle through a set of specific reactions, the attractor is a loop or orbit in phase space. Perturbations in such systems tend to relax back to a single state or cycle of states. An attractor is the state or set of states to which the system eventually evolves.

Systems exhibiting chaos do not have simple trajectories, yet may have attractors. In this case, the attractor is not a point or loop, but a region of space, and is called a strange attractor. The figure below shows an example of a strange attractor. This attractor is known as the Lorenz attractor because it arose in the model of the weather system

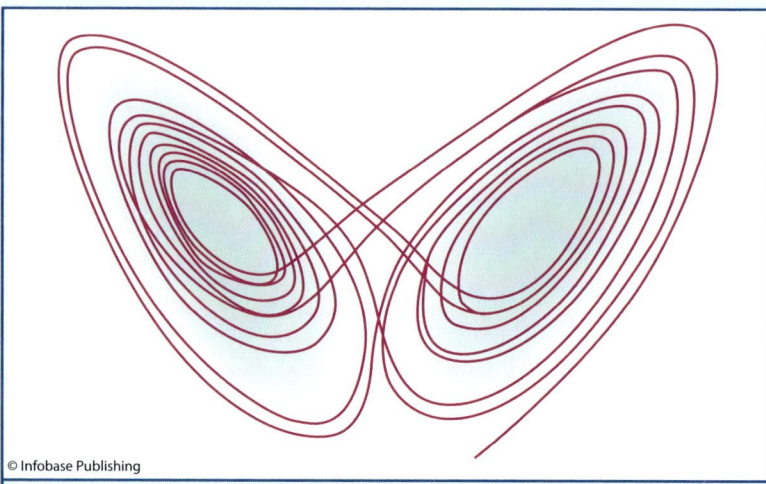

© Infobase Publishing

Attractors may have interesting shapes. This phase space diagram shows what is called the Lorenz attractor, the shape of which strikingly resembles the wings of a butterfly.

Computer-generated image of a fractal called a Mandelbrot fractal, named for pioneering French mathematician Benoit Mandelbrot *(Friedrich Saurer/Photo Researchers, Inc.)*

he was studying. The different trajectories remain in the same general area or areas, yet they do not precisely repeat, as a cyclical system would do. Strange attractors can have interesting patterns and a great deal of complexity. Physicists and other scientists study these attractors to learn about the dynamical behavior of the system and the interaction among its components.

Some of the patterns are fascinating mathematical structures in their own right. A *fractal* is an object that displays self-similarity—each part of it has a structure similar to the whole, rather like an A in which the lines are, upon microscopic examination, not fully dark but are made up of a dense cluster of tiny As, and so on. Structures that appear in the phase space of chaotic systems are often fractals—strange attractors, for example, are fractals, which means the pattern when viewed on a small scale will resemble the pattern of a larger scale.

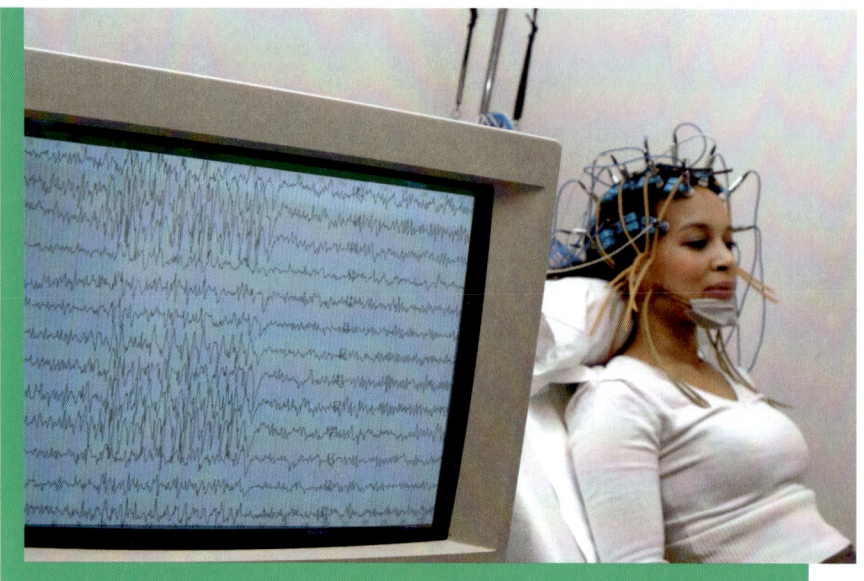

One of the ways researchers study brain activity is by recording signals from the surface of the scalp, a technique called electroencephalography. *(AJPhoto/Photo Researchers, Inc.)*

thoughts and behavior. But the development of chaos theory and nonlinear dynamics has given researchers who study the brain another valuable tool.

The human brain consists of hundreds of billions of electrically excitable cells called neurons. Neurons code and transmit information in the form of electrical impulses, the rate and/or timing of which carries the information. For example, light impinging on the retina—a layer of cells at the back of the eye—causes some of the neurons to increase their rate of impulses, signaling the presence of light. Neurons conduct impulses down a long projection called an axon. Transmission of information from neuron to neuron occurs across junctions known as synapses, usually formed when an axon of one neuron meets another cell. Large numbers of neurons communicate with one another via these junctions, forming neural networks that process information and, in a manner scientists have yet to comprehend, give rise to perception, thoughts, and motor commands to initiate action.

One of the most popular hypotheses of brain function is to liken the brain to a computer. The brain receives input from the senses, processes the data, stores relevant or noteworthy bits of information in memory, and emits an "output"—thoughts, perception, and behavior. Researchers have traced the flow of information in neural networks; for example, scientists who study visual perception have recorded the change in neural activity in a chain of connected networks as they process a visual stimulus, such as a geometrical figure or the face of a friend. The activity appears to code for a particular stimulus by breaking it down into its components—color and shape, for instance.

One of the problems with the computer hypothesis is that scientists do not understand how the brain unites these information streams into a perception, nor do scientists understand how the brain chooses which among the many neural networks should be attended to at a given time. No one knows how a familiar face, a specific noise, or a long-remembered smell of, say, mown grass suddenly captures a person's attention and elicits a train of associated memories.

A newer model of the brain views it as a dynamical system. In this hypothesis, the electrical activity of a group of neurons forms a (large) set of variables that traces trajectories in phase space. Because neurons and neural networks are nonlinear systems and often exhibit complicated, seemingly disordered trains of impulses, chaos theory might be applicable.

Walter J. Freeman, a biologist at the University of California, Berkeley, developed a model in the 1990s of the olfactory system—the sense of smell. This model incorporates chaotic dynamics and is mathematically sophisticated, but in simple terms patterns of activity that represent a specific smell are similar to attractors. The system "recognizes" odors when their activity falls into these patterns, and similar odors that resemble one another may fall into the same attractor basin, facilitating recognition. An advantage of this model is that new inputs can swiftly nudge it into another state; incoming information represents a perturbation that easily changes the system's dynamics, moving it into another state and perhaps another attractor. In this way, the system can process a number of stimuli rapidly. The computer hypothesis of the brain generally requires some sort of mysterious controller to manage and coordinate the processing of a series of stimuli.

But tremendous difficulties arise when researchers try to find evidence to support models of brain function based on dynamical systems and chaos theory. Thousands and, more often, millions of neurons are involved in any given function, and this vast number of cells generates too much activity for scientists to collect. At most, only a few dozen or perhaps 100 neurons can be recorded at the same time. Other measurements, such as electroencephalography, record the overall activity of thousands or millions of neurons, smeared into one signal; this technique is immensely useful for studying the behavior of neural networks, but the data are difficult to analyze in terms of what each neuron is doing.

Describing these data in terms of chaos theory is also problematic. In a 2003 review of chaos and brain function, Henri Korn and Philippe Faure of the Pasteur Institute in France wrote that chaotic models of neural networks involved in sensation and perception "continued to attract numerous researchers despite unconvincing experimental results (since there are no definite tests for chaos when it comes to analyzing multidimensional and fluctuating biological data)."

Yet researchers at the frontier of science are continuing to try to apply chaos theory to the activity of neurons and neural networks. In the case of single neurons, the data is more complete and manageable, and the dynamics of a cell can be chaotic. Researchers expect many "higher" systems will also prove so. Korn and Faure note that "although a convincing proof of chaos (as defined mathematically) has only been obtained at the level of axons, of single and coupled cells, convergent results can be interpreted as compatible with the notion that signals in the brain are distributed according to chaotic patterns at all levels of its various forms of hierarchy."

TURBULENCE, JET STREAMS, AND WEATHER

Other systems are already well known to be chaotic. Weather is the earliest and one of the most prominent phenomena in which researchers have studied chaos. Scientists who study weather are also interested in the flow of fluids—the term *fluid* can refer to air or liquid. The atmosphere consists of air that readily moves, carrying heat and equalizing pressure with winds. Wind is an important element in weather.

Disorderly motion in a fluid is called *turbulence.* The figure on page 129 shows an example, contrasting turbulence with smooth flow, called laminar flow. Turbulence often occurs when a high-speed object such as a car races through the air, leaving behind a complicated wake. Another example is the mixing of cold and hot air, creating complex wind pat-

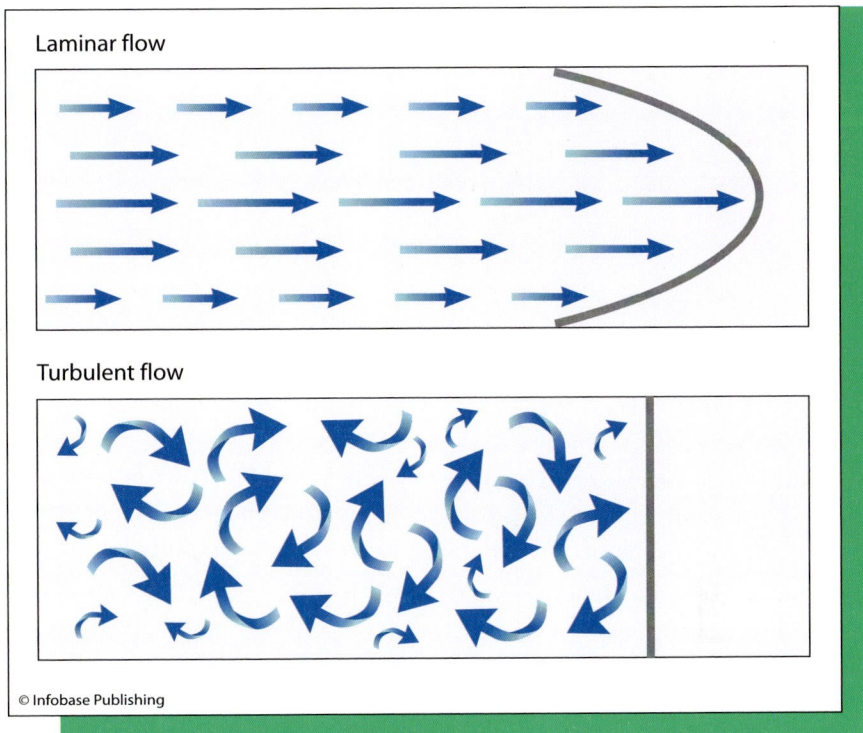

Laminar flow

Turbulent flow

© Infobase Publishing

In laminar flow, shown on top, the motion is smooth and even, as indicated by the arrows. Turbulent flow, shown on bottom, consists of particles moving in a variety of directions.

terns as the temperature and pressure equalize. The flight of an airplane can be severely disrupted when it encounters disorderly wind patterns called clear air turbulence—turbulence that is impossible to see because it does not involve smoke or clouds. An airplane that suddenly hits a patch of rough air can make an abrupt transition from a smooth ride to a rocky one, tossing about any loose items (such as passengers who are not wearing seat belts).

Scientists are not sure exactly how often and in which cases chaos theory applies to turbulence. A disorderly flow does not necessarily mean that it exhibits chaos, since it could instead be due to random motion—"chaotic" only in the sense of being stochastic and disorganized. Recall that chaos theory refers to a deterministic system, the motion of which appears disorderly but actually contains a great deal of order (as revealed in phase space diagrams, for example).

NASA researchers visualize the airflow around this model jet with smoke illuminated by a laser light sheet. *(NASA/ Langley Research Center)*

Yet as Lorenz discovered during his computer simulation, weather and the atmosphere are known to have chaotic dynamics in a lot of cases. The butterfly effect does not give weather forecasters much reason for optimism, but the atmosphere does contain some order amid the unpredictability. Extremely important components of the weather include narrow bands of high-speed wind in the upper atmosphere known as jet streams. The sidebar on page 131 provides additional information about these winds.

Jet streams form because of temperature differences in air masses—for instance, when a mass of polar air meets warm tropical air—which result in pressure gradients, causing wind to arise. Earth's rotation tends to wrap the jet streams around the globe. But how do these narrow bands of wind stay coherent? Logic would suggest that the turbulent, disorganized motion of the atmosphere would break them up, yet jet streams maintain their tight channels of flow for long periods of time. They are "rivers" of orderly motion in a turbulent sea, and scientists do not understand exactly what maintains them. Chaos theory may be playing an important role in these important systems.

CHAOS AND QUANTUM MECHANICS

Chaos theory helps scientists find order in systems whose behavior seems at first glance to be thoroughly disordered, but this is because systems to which chaos theory applies are deterministic—equations and formulas determine each step in their evolution. These systems are sensitive to initial conditions, which renders long-term predictions difficult or even next to impossible, but the systems are not purely random; their motion from point to point in phase space is not based solely on chance.

Jet Streams

Jet streams form at altitudes above 20,000 feet (6,100 m) and can stretch across large portions of the globe, such as the entire United States. The streams are usually around 200 miles (320 km) wide and less than 3 miles (4.8 km) deep. Speeds vary, but at the core of the stream winds may blow up to about 300 miles/hour (480 km/hr) during winter, when jet streams are strongest. The location of the jet stream shifts, depending on the movement of warm and cold air masses.

Several jet streams normally blow in the Northern Hemisphere. The northernmost stream, known as the polar jet stream, tends to snake its way over the United States in winter, with the winds generally blowing from west to east. Weather forecasters often show the location of the polar jet stream (which they usually refer to as "the" jet stream) because it affects storms and weather systems and reflects temperature boundaries.

Winds such as jet streams arise when air under high pressure flows toward regions of lower pressure. (Pressure differences arise, for example, when warm air rises, creating a zone of low pressure; as the temperature falls, air descends, creating high-pressure zones.) Air in the upper atmosphere is thinner, and temperature differences create relatively large pressure differences, so jet streams tend to be strong. Jet streams can be quite long, so they tend to curve, circling the globe, due to the Coriolis effect, named after the French mathematician Gaspard-Gustave Coriolis (1792–1843). The Coriolis effect occurs because Earth is a rotating sphere (roughly), and the speed of the surface depends on latitude. (For instance, the motion of a rotating sphere at the equator is fast, while the poles—positioned at the axis of rotation—scarcely spin around at all.) This speed variation makes the motion of wind or flying objects appear to curve with respect to the surface. Because of this effect, jet streams blow from west to east.

The deterministic nature of chaotic systems contrasts with the tenets of one of the main foundations of modern physics—quantum mechanics. The equations of quantum mechanics, which governs the behavior of atoms and molecules, are not deterministic but instead give only probabilities. For example, quantum mechanics may dictate that a particle has a 60 percent chance of evolving into state A and a 40 percent chance of evolving into state B, but it does not specify which one will be true for any given instance. Quantum mechanical probabilities are only noticeable on the microscopic level of atoms and molecules, though many people assume that quantum mechanics holds true for all systems. "Macroscopic" systems are deterministic—or seem that way—because they contain a huge number of atoms and molecules and the fluctuations due to probabilities of single particles do not matter because it is the group that governs the system's behavior. The group's behavior is predictable in principle, even though individual atoms and molecules behave stochastically (their behavior is due to chance). In the example described above, 60 percent of the particles will end up in state A and 40 percent in B, and although quantum mechanics does not specify which particles are in A and which are in B, the percentages determine a macroscopic system's behavior, which is predictable given these percentages.

Another contrast between chaos theory and quantum mechanics concerns the equations that describe the evolution of states. The principal equation used in quantum mechanical calculations is the Schrödinger equation, discovered by the Austrian physicist Erwin Schrödinger (1887–1961). This equation is generally linear.

Stochasticity and linearity in quantum mechanics would seem to rule out any application of chaos theory. But researchers are finding behavior similar to chaos on microscopic scales.

In 2008, Brian Saam, a physicist at the University of Utah, and his colleagues observed unexpected order in a quantum mechanical system. Saam and his team studied the state of a group of billions of xenon atoms, frozen in a solid at -321°F (-196°C). The researchers focused on a property called spin, which as described in chapter 2 is a quantum mechanical property. These xenon atoms can have a spin "direction" that is either up or down; the specific direction depends on interactions between the atoms in the solid and any external fields or impinging radiation. In the initial portion of the experiment, the researchers made four xenon crystals (composed of billions of atoms) and aligned the spins of each sample with the aid of pulses of radio waves. But each sample received a different radio wave pulse, so the arrangement of spins in the samples differed. The researchers then used a sensitive technique called nuclear magnetic resonance to observe how the spins evolved.

Despite the atoms being frozen in place, the spins could change direction as they interacted with one another. Each sample rapidly changed from its original configuration in a complicated fashion. But even though the original configuration was different, their behavior was very nearly the same over the long term (which, for this quickly changing property, is a few milliseconds). In a news release posted on August 9, 2008, at Science*Daily,* Saam said, "Somehow despite the fact these spins have very complicated interactions with each other and started out in completely different orientations, they end up all moving in the same way after several milliseconds." He noted, "That's never been seen before in a quantum mechanical system. These guys are dancing together."

What Saam and his coworkers found was an unexpected order in a complicated system. This situation strongly resembles chaos.

Physicists working on experiments such as this one often refer to the phenomenon as "quantum chaos." This term is somewhat misleading since quantum mechanics and chaos theory have profound incompatibilities, and one of the primary characteristics of chaos—sensitivity to initial conditions—is not observed in quantum mechanics. Yet Saam and other researchers have shown that orderly chaoslike properties can emerge in systems governed by quantum mechanics. This leads to interesting questions regarding exactly what sort of phenomenon is occurring in quantum mechanical systems and its relationship to "classical" chaos theory.

Researchers at the frontiers of physics are studying possible relationships between chaos theory and quantum mechanics, but no firm solution has been found. An exciting possibility is that a new set of principles in physics could arise from these studies. As Saam noted in the news release of August 9, 2008, "When you look at all the technology governed by quantum physics, it's not unreasonable to assume that if one can apply chaos theory in a meaningful way to quantum systems, that will provide new insights, new technology, new solutions to problems not yet known."

CONTROLLING CHAOS

The discovery of chaos theory has led to an increased understanding and a new perspective of nonlinear systems, whose behavior had earlier appeared completely random and disordered. Now that scientists have glimpsed some order in chaotic systems, people have begun to wonder if they can use this knowledge to manipulate or control the system's behavior.

Phase space trajectories of a chaotic system often move toward an attractor, although unlike simple systems, this region of phase space is not a fixed point or a single loop in which the variables are confined to a limited set of values. Within this region the system may stay for a while, assuming a variety of states during its unstable orbit, until a perturbation throws the system out of this region and toward another. This behavior differs from a totally random or stochastic process, which jumps from state to state by chance. Korn and Faure, in their 2003 paper, noted "the possible benefits of chaotic systems over stochastic processes, namely of the possibility to control the former. Theoretically such a control can be achieved by taking advantage of the sensitivity of chaotic trajectories to initial conditions and to 'redirect them,' with a small perturbation, along a selected unstable periodic orbit, toward a desired state."

The advantage of this technique is the ability to nudge a system into a certain region of its phase space, perhaps keeping it there for a while. For example, if a system's performance is best in a certain range of values, operators will want to prolong the time a system stays in these states. Such techniques to "control chaos" rely on the sensitivity of these nonlinear systems. But care must be taken when applying perturbations to a dynamical system, since a substantial force could modify the whole dynamics of the system. Perturbations involve fine-tuning and deft touches rather than a heavy-handed approach.

In 1990, the University of Maryland researchers Edward Ott, Celso Grebogi, and James A. Yorke published a paper, "Controlling Chaos," in *Physical Review Letters*. The researchers described a technique that maintains a chaotic system in a periodic pattern, with the variables cycling through a number of values—in phase space, this means a periodic orbit. Because chaotic systems are unstable, a tiny push or perturbation is needed once per cycle of the orbit in order to keep it in place. The technique calculates the amount of perturbation needed by using feedback;

measurements of the system reveal the difference between the system's present state and the desired state, and the controller adjusts one or more of the variables to reduce the difference. This continual feedback control keeps the system from straying from its periodic orbit.

Although the technique of Ott, Grebogi, and Yorke can be effective, it only works for systems with few variables, or in other words low-dimensional systems. High-dimensional systems are so unstable that they wildly career into new orbits; unfortunately for researchers who study the control of chaos, most real-world systems that one would want to control are high dimensional. Attempts to control high-dimensional systems have not met with a great deal of success, but improvement in this area would bring impressive advances. For example, runaway electrical activity in the brain known as seizures may be brought under control with such techniques. Heart arrhythmias are another set of problems that might be amenable to control.

Research on chaos and dynamical systems in general has many other potential benefits. Chaotic systems are a type of complex system. A dynamical system may be complex because it has a huge number of interacting parts, or it may be complex due to the complicated ways that its components interact, but in either case such systems are difficult to study and understand. Scientists often analyze objects under study by breaking them into parts, but complex systems often defy such analysis because their properties emerge from the collective behavior of the components. Studying a component or piece of the system in isolation fails to reveal much about the way the system as a whole works.

One prominent example of a complex system is Earth's climate. A huge number of factors, including winds, temperatures, precipitation, ocean currents, topography, and biological organisms affect the behavior of this system. Observations over the past century indicate that average surface temperatures are rising, and no one is certain how and to what extent this change may ripple through the system. The importance of this system to human society makes a study of this system imperative, despite its notorious complexity. Chaos theory offers an important method of finding order in certain complex systems and in some cases controlling it, which may apply in this case. And perhaps the further study of dynamical systems may uncover more ideas similar to chaos theory.

CONCLUSION

The name *chaos theory* seems ironic considering that scientists have begun using it to find and in some cases introduce more order in certain dynamical systems. Although systems that exhibit chaos are extremely sensitive to initial conditions, which as Lorenz discovered makes them difficult to predict, they do not haphazardly jump from state to state.

An important application of complex systems in physics recently arose in the development of nuclear fusion reactors. As described in chapter 1, nuclear fusion is a reaction in which nuclei fuse, or join, releasing a large amount of energy in the process. Fusion is the mechanism that makes the Sun and other stars shine and offers the potential of clean, plentiful energy for society if an economically viable reactor can be built. The problem is that the necessary conditions for fusion to occur involve exceptionally high temperature and pressure, which exist in the center of a star but are expensive to recreate on Earth's surface. Researchers have already built devices capable of producing fusion, but the cost of creating the necessary conditions exceeds the value of the energy output.

To generate fusion reactions, physicists generally must confine the material under extremes in temperature and pressure, but such conditions will melt or explode containers. One method of confining the material involves the use of magnetic fields, which can exert constraining forces on a material, such as a plasma (a hot gas consisting of ions). This method, known as magnetic confinement, is the choice of several research programs, including a large, multibillion dollar international project called ITER.

One of the serious difficulties facing magnetic confinement projects such as ITER is that the hot plasma can occasionally spurt through the magnetic field. Bursts are brief, but cause discharges similar to lightning, which severely damage reactor components. Replacement and maintenance expenses add to the cost of running a fusion reaction, reducing the possibility of developing an economically viable reactor.

Abrupt transitions in behavior such as discharges are common in complex systems. Todd Evans, a physicist at General Atomics, a company based in San Diego, recently proposed a possible solution that involves small perturbations in the system. In this technique, a small coil generates a magnetic field that perturbs the reactor's main field. The science journalist Geoff Brumfiel wrote an article about this idea in 2006 at Nature. com. Brumfiel noted that the perturbation "weakens the field just enough to let a little bit of plasma leak out through the bottom, relieving some of the pressure in the system and preventing it from bursting."

But as is the case with many complex systems, physicists do not understand exactly how this technique works or whether it should be incorporated into ITER. Brumfiel quoted Philippe Ghendrih, a physicist at the French Atomic Energy Commission: "This area of the machine is far too complicated to match the simple theoretical ideas we have been working with." Ghendrih believes that further study will be required.

Any technique that helps enable nuclear fusion technology would be worthwhile. Unlike fission reactions, which are the basis for all nuclear reactors operating today, fusion generates few dangerous emissions or radioactive residues, and the hydrogen isotopes used as fuel are cheap and plentiful. Fusion would be an excellent power source if the problems of plasma confinement can be overcome. The study of chaos theory and complex systems, along with other branches of physics, will be a major contributor in this effort.

Many other complex systems also offer substantial but potentially rewarding challenges. At first glance, these systems may appear far too disorderly to ever be comprehended. But sometimes patterns emerge. The mathematical notions of chaos theory have helped scientists to discover order where none was seen before. As this frontier of science expands, perhaps many other complex systems will prove to be a lot less random than they seem to be.

CHRONOLOGY

1890	While working on the three-body problem, the French mathematician Henri Poincaré (1854–1912) discovers some of the principles of the mathematical theory of chaos.
1901	The American mathematician Josiah Willard Gibbs (1838–1903) develops the idea of phase space.
1906	Poincaré conjectures that some systems are so sensitive to initial conditions that they are difficult or impossible to predict.

1961	Edward Lorenz (1917–2008) observes a computer model that shows extreme sensitivity to initial conditions—a small change in the initial conditions causes a drastic change in the model's output.
1963	Lorenz publishes his observations in a paper, "Deterministic Nonperiodic Flow," in the *Journal of the Atmospheric Sciences.*
1971	The Belgium physicist David Ruelle and the Dutch mathematician Floris Takens describe strange attractors.
1972	Lorenz gives a talk titled "Predictability: Does the Flap of a Butterfly's Wings in Brazil Set off a Tornado in Texas?" discussing the butterfly effect.
1975	The University of Maryland mathematician James A. Yorke and his colleague Tien-Yien Li begin using the term *chaos* with respect to dynamical systems.

The French mathematician Benoît Mandelbrot introduces the term *fractal* to describe self-similar structures. |
1977	The first scientific conference on chaos theory takes place in Como, Italy.
1988	James Gleick publishes *Chaos: Making a New Science,* popularizing chaos theory.
1989	The Boston University mathematician Robert L. Devaney publishes the book *An Introduction to Chaotic Dynamical Systems,* which provides a rigorous mathematical definition of chaos.
1990	The University of Maryland researchers Edward Ott, Celso Grebogi, and James A. Yorke publish one of the earliest techniques to "control chaos."

| 2001 | The University of Texas researcher Mark G. Raizen and his colleagues discover that chaos can enhance a quantum mechanical phenomenon known as tunneling (movement across a barrier). |

2001 The University of Texas researcher Mark G. Raizen and his colleagues discover that chaos can enhance a quantum mechanical phenomenon known as tunneling (movement across a barrier).

2005 Gernot Stania and Herbert Walther, researchers at the Max Planck Institute of Quantum Optics in Germany, demonstrate chaoslike behavior in atoms.

2006 Todd Evans of General Atomics, a company in San Diego, develops a possible method of eliminating burst discharges in magnetic confinement fusion.

2008 The University of Utah physicist Brian Saam and his colleagues observe a surprising amount of order in a quantum mechanical system containing large numbers of xenon atoms.

FURTHER RESOURCES

Print and Internet

Brumfiel, Geoff. "Chaos Could Keep Fusion under Control." Nature. com (5/22/06). Brumfiel briefly describes the use of magnetic perturbations to prevent damaging discharges from occurring in nuclear fusion reactors.

Emanuel, Kerry. "Edward N. Lorenz (1917–2008)." *Science* 320 (5/23/08): 1,025. This one-page article summarizes the life and career of this pioneering scientist.

Gleick, James. *Chaos: Making a New Science.* New York: Penguin, 1988. In one of the first popular accounts of chaos theory, the science writer James Gleick describes the early findings and the researchers who uncovered them.

Korn, Henri, and Philippe Faure. "Is There Chaos in the Brain? II. Experimental Evidence and Related Models." *Comptes Rendus Biologies* 326 (2003): 787-840. The researchers offer a thorough review of experiments and models of chaos in the brain.

Lorenz, Edward N. *The Essence of Chaos.* Seattle: University of Washington Press, 1993. Written by one of the discoverers of chaos, this book discusses the development of the theory and its meaning and applications.

Massachusetts Institute of Technology. "Edward Lorenz, Father of Chaos Theory and Butterfly Effect, Dies at 90." News release (4/16/08). Available online. URL: http://web.mit.edu/newsoffice/2008/obit-lorenz-0416.html. Accessed June 22, 2009. This news release marks the passing of Edward Lorenz, a longtime researcher at MIT.

Ott, Edward, Celso Grebogi, and James A. Yorke. "Controlling Chaos." *Physical Review Letters* 64 (1990): 1,196–1,199. In this important paper, Ott and his colleagues introduce a method of controlling chaos in a low-dimensional system.

Peitgen, Heinz-Otto, Hartmut Jürgens, and Dietmar Saupe. *Chaos and Fractals: New Frontiers of Science.* New York: Springer-Verlag, 1992. This book describes the subject with the help of many illustrations and examples. Although the authors do not shy away from the mathematical aspects of chaos and fractals, the examples are simple and adroitly explained.

Science*Daily*. "Quantum Chaos Unveiled?" News release (8/9/08). Available online. URL: http://www.sciencedaily.com/releases/2008/08/080806140211.htm. Accessed June 22, 2009. The University of Utah researcher Brian Saam and his colleagues conduct experiments in which a frozen solid of xenon atoms displays behavior remarkably similar to chaos.

Williams, Garnett P. *Chaos Theory Tamed.* Washington, D.C.: Joseph Henry Press, 1997. Williams provides a thorough and accessible introduction to chaos theory.

6

STRING THEORY AND THE FOUNDATIONS OF PHYSICS

People have long wondered what sorts of substances compose the world at its most basic level and how these substances interact. The ancient Greek philosopher Empedocles (ca. 490–430 B.C.E.) believed that there were four fundamental substances, or elements—earth, fire, air, and water. Much later, scientists such as the British chemist John Dalton (1766–1844) proposed the existence of atoms. Physicists of the 20th century discovered a large number of different particles, as described in chapter 2, and established the standard model, which describes the elementary particles and the forces with which they interact.

Yet the story is not finished. Gravitation is not important in particle interactions, at least not at the energies at which particle accelerators operate and has not been fully incorporated into particle physics theory. The standard model may not be the whole story because it does not explain all the forces of nature—scientists are not sure if the model applies to gravity—and physicists are searching for a comprehensive theory. There are many possibilities. One possibility that has gotten much attention recently is string theory, in which the fundamental substances of the universe are thin, vibrating strings.

String theory involves a lot of advanced mathematics. These techniques will not be described in this chapter, but they offer physicists elegant methods with which to unify different concepts. One theory to explain all observations is much more satisfying than many disparate theories, which explain one or just a few observations and may even contradict one another.

But theories must be supported by experimental evidence, and ambitious theories such as string theory will not instill much confidence until researchers successfully conduct experimental tests. Albert Einstein theoretically deduced a surprising relationship between energy mass, for example, to which numerous experiments and observations have been found in agreement (see chapter 1), and the astonishing implications of Einstein's relativity theory have also received much support—the British astronomer Sir Arthur Stanley Eddington (1882–1944) even traveled to the African island of Príncipe in 1919 to make observations of a solar eclipse that helped decide whether Einstein was right.

Experiment and theory go hand-in-hand—one without the other is unsatisfactory. Chapter 4 described superconductors, a topic in which researchers have an abundance of experimental data and are presently in search for a theory to explain them. The discussion in this chapter belongs to the opposite case, in which researchers have a theory and are in search of experimental data. This chapter describes the basic concepts of string theory and how physicists are seeking methods of testing it.

INTRODUCTION

The allure of unifying theories is strong. Consider the work of Sir Isaac Newton on gravitation. When he formulated the universal law of gravitation, Newton accounted for seemingly different motions such as the fall of an apple on Earth's surface and the orbit of a planet, all within a single unifying concept. Another prominent unification in physics occurred when the Scottish physicist James Clerk Maxwell (1831–79) brought electricity and magnetism together in his equations of electromagnetism. Particle physicists of the 20th century expended a great deal of effort to boil down the many particles and interactions into the standard model.

More can be done. For example, no one knows why there are four forces—electromagnetism, strong nuclear force, weak nuclear force, and gravitation. Perhaps these forces are manifestations of a single underlying concept. As yet unfinished theories, known generally as grand unifying theories, attempt to merge electromagnetism, strong nuclear force, and weak nuclear force into one. An even bolder attempt, called the theory of everything, aims to explain all four. String theory is a candidate for this theory of everything.

But wrapping all of the fundamental concepts of physics into one neat package is not going to be easy. In particular, two of the staunchest foundations of physics—quantum mechanics and Einstein's general theory of relativity—do not mesh well at all.

Quantum mechanics governs the behavior of small particles, which do not obey the equations of classical physics as formulated by Newton. Chapter 5 described chaos, in which order can be found in the evolution of certain dynamical systems, and contrasted this phenomenon with quantum mechanics. Unlike systems that can exhibit chaos, quantum mechanics does not ascribe definite motion or activity to a specific particle. Instead, solutions to the equations of quantum mechanics provide a probability that a certain particle or system will evolve to a given state. Quantum mechanics also has an inherent amount of uncertainty. Heisenberg's uncertainty principle—named after its discoverer, the

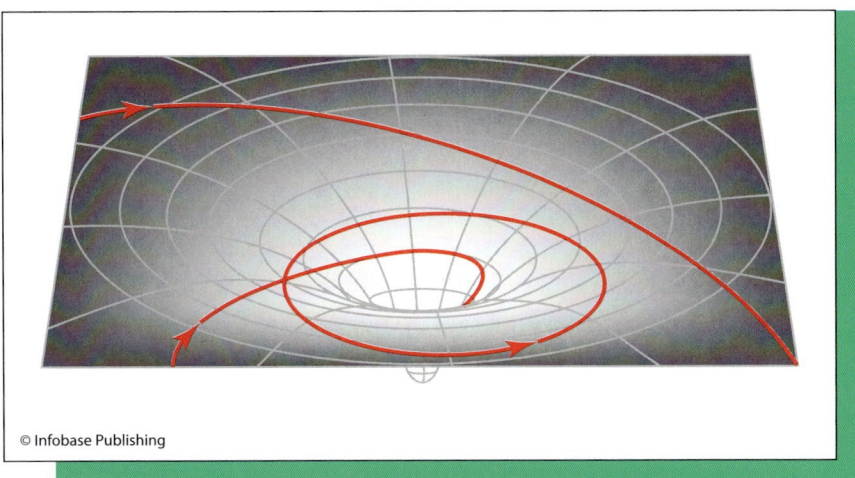

© Infobase Publishing

The sheet in this figure represents space-time, which is warped by the presence of a massive object in the center. Other bodies roll toward the object. Their motion depicts the manner by which the massive object in the center gravitationally attracts other masses.

German physicist Werner Heisenberg (1901–76)—states that in quantum mechanics, certain pairs of measurements such as the position and momentum of a particle can never be made simultaneously with perfect precision.

The probabilistic and uncertain nature of quantum mechanics differs from Einstein's general theory of relativity. This theory describes gravity in terms of the curvature of space and time, or *space-time*; whereas Newton envisioned gravitation as an attractive force acting between two bodies, Einstein thought that bodies warp or curve space-time, causing other bodies to fall toward them, as illustrated in the figure on page 143. Although physicists still apply Newton's simpler universal law of gravitation in many situations, Einstein's theory is more accurate and generally applicable. *General relativity* is a deterministic theory, which means that the equations determine the behavior of the system rather than specifying the probability that it will evolve one way or another.

Much experimental evidence supports both quantum mechanics and general relativity. Both theories accurately describe different realms of physics—quantum mechanics governs the behavior of atoms and molecules, and general relativity depicts gravitation, which, being the weakest of the four forces, becomes important only when large masses are present. But at some point physicists would like to put the two together, reconciling their differences. It seems odd that nature would have two incongruous laws.

Formulating some kind of *quantum gravity*—a melding of the ideas of quantum mechanics and general relativity—has not been successful. In a 2002 article in *Astronomy,* Edward Witten, a theoretical physicist at the Institute for Advanced Study in New Jersey, wrote that "direct attempts to express general relativity in quantum mechanical terms have led to a web of contradictions, basically because the nonlinear mathematics Einstein used to describe the curvature of space-time clashes with the delicate requirements of quantum mechanics."

The trouble physicists have had in their efforts to unify quantum mechanics and general relativity is one of the reasons why they are eager to search for new ideas. String theory has generated a lot of excitement because it has the potential to bridge the gap between quantum mechanics and general relativity. If string theory forms the basis for quantum gravity, it will have reconciled these two pillars of physics into one elegant, unified concept.

String theory is actually not a single theory but a family of theories that use the same concept of strings as fundamental components of matter. In 1974, a California Institute of Technology researcher John H. Schwarz and his colleague Joel Scherk discovered a form of string theory in which one of the strings had all the right properties to be a graviton—a hypothetical particle that mediates the force of gravitation. In the standard model, special particles mediate or "carry" each of the four forces; for example, the electromagnetic force results when objects exchange photons, the carrier of the electromagnetic force. According to this theory, gravitation also needs a mediating particle, dubbed the graviton, which would have certain properties. The graviton is virtually impossible to spot in particle accelerator experiments because gravitation is much weaker than the other forces, but it appeared in Schwarz's version of string theory.

New ideas about space and time also emerged in string theory. These concepts were able to blend smoothly into general relativity, connecting the stochastic nature of quantum mechanics with the geometry of Einstein's space-time. As Witten wrote in his 2002 article, "Einstein based his theory of gravity on his ideas about space-time, so any theory that modifies Einstein's gravitational theory to reconcile it with quantum mechanics has to incorporate a new concept of space-time. String theory actually imparts a 'fuzziness' to all our familiar notions of space and time, just as Heisenberg's uncertainty principle imparts a basic fuzziness to classical ideas about the motion of particles."

The "fuzziness" arises because of the strings. Einstein's general relativity depicts the force of gravitation as a smooth, clockwork operation, but quantum mechanics indicates that at extremely small scales of space and time, objects behave indeterminately, jumping from state to state or existing in a jumble of different states. Quantum mechanics eliminates any notion of smoothness at the atomic level. In string theory, strings can stretch across the choppiness of quantum scales, which allows them to fit into the framework of gravitation theory without ignoring the reality of quantum mechanics.

VIBRATING STRINGS

The standard model describes electrons and quarks as pointlike particles with no internal structure, and there is no evidence that particles such as electrons have any spatial extent at all. A true point particle

Planck Length

Planck length involves a specific group of constants. A constant in physics is a value that does not change, and physicists usually represent them with specific symbols. For example, the speed of light in a vacuum is a constant, represented by c (which is the first letter of the Latin word *celeritas*, meaning swift). Sometimes constants arise when scientists discover that one variable is proportional to another, which means for variables x and y, the equation can be written $y = kx$, where k is a constant known as the constant of proportionality. Newton discovered that the force of gravitation is proportional to the square of the masses and inversely proportional to the square of the distance between them. Physicists often represent this gravitational constant as G.

The unit of length called Planck length, l_p, is given by a formula involving three constants—the speed of light, the gravitational constant, and Planck's constant, $\hbar$. Planck's constant is an important number in quantum mechanics and

would have zero dimensions. String theory replaces this zero-dimensional concept with a string; objects such as electrons and quarks are not particles but rather one-dimensional strings. In this idea, strings are visualized as a line in geometry—a one-dimensional object having length but no width. Although these abstractions are difficult to conceptualize—familiar objects are three-dimensional—objects or phenomena in nature need not conform to human perception.

If strings are to account for particles such as the graviton, Schwarz and his colleagues calculated that the strings must be under an exceptional amount of tension—a force that keeps them taut. This force causes a loop of string to contract into an incredibly tiny size of roughly 0.4×10^{-33} inches (1×10^{-33} cm), a magnitude of length called Planck length, named after one of the pioneers of quantum mechanics, the German appears in many equations. The formula for Planck length is as follows:

$$l_p = \sqrt{\frac{G\hbar}{c^3}}$$

Since this length involves constants that are critical in gravitation and relativity as well as quantum mechanics, it is reasonable to suppose that it could play a vital role in any form of quantum gravity. This view strengthened string theory's claim on quantum gravity when Schwarz and Scherk calculated the length of strings to be a similar value.

The magnitude of Planck length also marks what many physicists believe is the edge of observability. Consider the particle accelerators described in chapter 2, which send particles hurtling at high speeds into a target. Accelerator experiments need the tremendous energy of collisions to probe the structure of tiny particles. Small distances require higher energy, but at some juncture, the energy is so great that it causes instability, possibly even a black hole. The distance at which this is likely to occur is the Planck length.

physicist Max Planck (1858–1947). Planck length is a critical distance, as discussed in the sidebar on page 146.

A string under tension can vibrate. Guitar strings, for example, stretch between two points. Strings vibrate at certain frequencies, as illustrated in the figure on page 148. Although the strings in string theory are not fixed at each end, they can also vibrate in characteristic modes or frequencies.

Differences in a string's properties result in different modes of oscillation. For a guitar string, the fundamental frequency depends on the string's length. In string theory, physicists have proposed that different kinds of oscillation correspond to different particles, so that a string vibrating in a certain way corresponds to a particle of given mass and electric charge. The strings can split or combine, resulting in processes

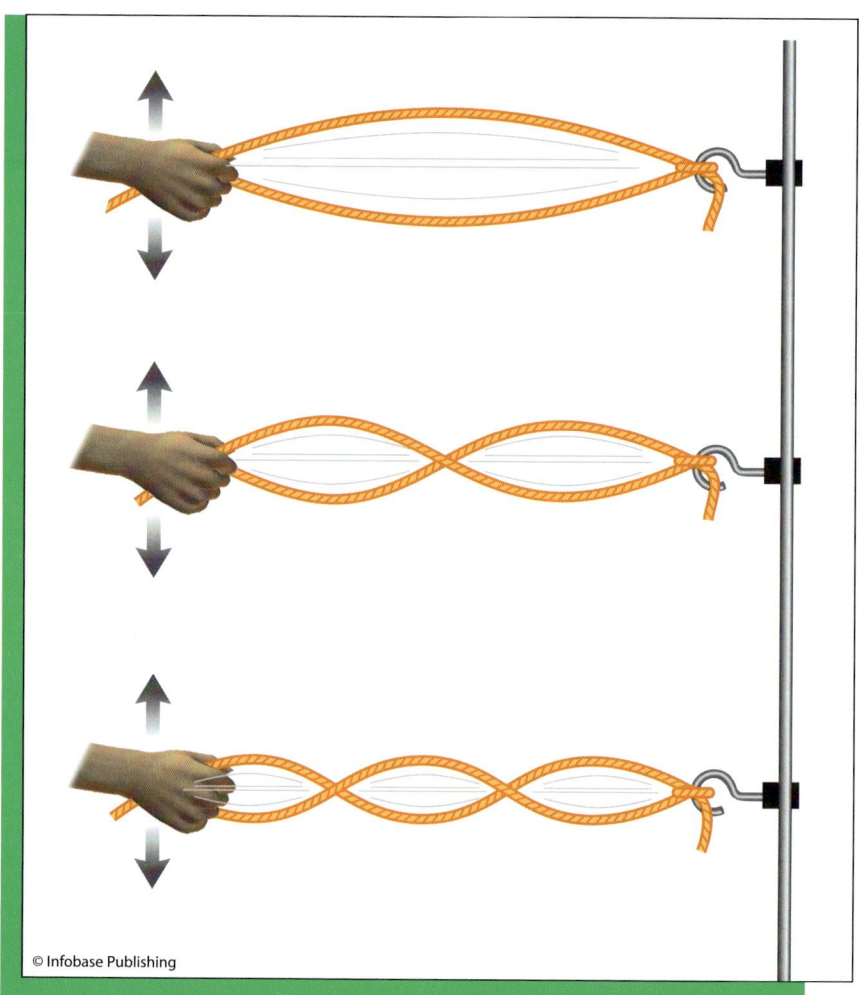

© Infobase Publishing

Strings vibrate at certain frequencies related to the length of the string. The longest wavelength, corresponding to the lowest frequency, is shown at the top of the figure. A string can also vibrate at half that wavelength—or twice the frequency—as shown in the middle, and a third of that wavelength—three times the frequency—as shown at the bottom, and so on. In general, the modes of vibration are multiples of the lowest frequency.

that emit "particles" or absorb them, which accounts for interactions in particle physics. Witten noted in his 2002 article, "In string theory, different harmonics correspond to different elementary particles. If string theory proves correct, all elementary particles—electrons, photons, neutrinos, quarks, and the rest—owe their existence to subtle differences in the vibrations of strings. The theory offers a way to unite disparate particles because they are, in essence, different manifestations of the same basic string."

Strings may be open or closed. If closed, they form a loop, whereas open strings have two distinct ends. Open and closed strings have different properties, which may play a role in determining what sort of vibration corresponds to each particle.

But the strings of string theory should not be confused with any kind of familiar object. An analogy with strings such as threads or guitar strings is useful, but string theory deals with objects that are much more mathematically abstract. Mathematical consistency of the theory is of primary importance, for if the theory contains paradoxes or contradictions, it cannot be a correct description of nature.

The mathematics of string theory is complex. In 1984, Schwarz and Michael Green, a researcher at the University of Cambridge in the United Kingdom, discovered a mathematical way of formulating the equations of string theory in a consistent way. There were several possible formulations of string theory, each of which seemed able to account for the physics of elementary particles and their interactions. This was a critical advance in string theory and sparked serious efforts on the part of mathematicians and theoretical physicists. In a 2004 *Science* article, the journalist Adrian Cho quoted Schwarz as saying, "Almost overnight, hundreds of people started working on this stuff."

But some peculiarities arose. String theory seemed to apply to a universe in which space-time has more dimensions than the four—three of space and one of time—to which scientists are accustomed.

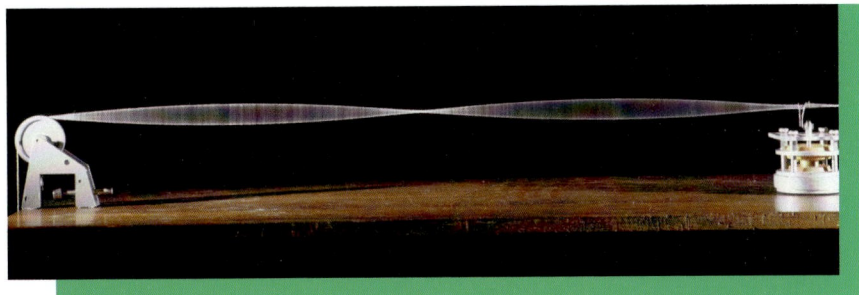

A string vibrating at twice the lowest frequency—note the stationary point in the middle *(Andrew Lambert Photography/Photo Researchers, Inc.)*

ADDED DIMENSIONS

Geometry often deals with objects with special dimensions, such as a one-dimensional line or a two-dimensional plane. People are used to living in three spatial dimensions, so it is difficult to imagine any other possibilities. In 1884, Edwin A. Abbott published a short book titled *Flatland: A Romance of Many Dimensions* (under the pseudonym A Square) that described creatures in a two-dimensional world. Abbott wrote the story partly to make fun of the society and culture of his time, but the book's popularity has endured for many reasons, especially its interesting presentation of what life would be like to an organism living in a world much different than the familiar three-dimensional one.

Mathematicians have little trouble handling multiple dimensions. Formulas in three-dimensional space may use representations such as (x, y, z), which represent a point in space on the x-, y-, and z-axis of a three-dimensional coordinate system, but these formulas can easily be extended to any dimensions. For example, a point in five-dimensional space can be represented as $(x_1, x_2, x_3, x_4, x_5)$. People cannot imagine or perceive such spaces, and artists cannot draw illustrations in them, but mathematicians can adapt formulas and systems with little trouble.

String theory was a bold new idea, but in order to be useful and accurate it must not contradict observation and experimental evidence. To incorporate the matter and energy concepts that particle physics and quantum mechanics have already established, the mathematics of string theory requires more than the usual four dimensions of space-time.

How many dimensions are needed depends on the theory. Some versions of string theory have a space-time of 26 dimensions, but these versions, while they can be made logically consistent, do not represent all of the known particles and physics. Other theories offer a rich depiction of the universe and its particles and manage to do so with 10 or 11 dimensions, as described in the following section. These are the versions of string theory that the rest of the chapter will discuss.

If space-time is really 10- or 11-dimensional, why do people perceive only four—three of space and one of time? Perhaps human perception is limited. Imagine, for instance, a flat bug crawling on a ball. If the bug is limited to the surface of the sphere, it can only know two dimensions, since it never encounters anything vertical. No notion of up and down would ever appear, even though the third dimension is actually present. The creature's sensation and perception would be limited but not the world in which it lived.

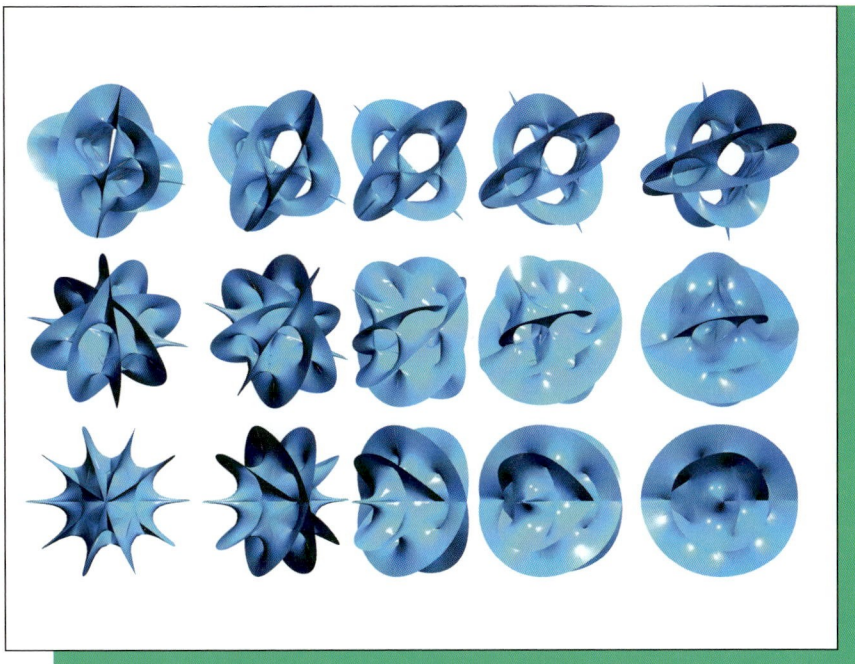

Examples of Calabi-Yau manifolds *(Laguna Design/Photo Researchers, Inc.)*

A similar situation may hold true for people. The extra dimensions of space-time required for a consistent, comprehensive string theory may be out of sensory reach for human beings. These dimensions could be "curled up," hidden or obscure. But the mathematics of string theory works, at least when these extra dimensions are available. In mathematics, the dimensions provide some wiggle room—extra variables with which theorists can treat the presently known laws of physics.

Extra dimensions in string theory may have a specific shape, which helps determine how the strings move and oscillate. A common shape used for this purpose is called a Calabi-Yau manifold, named after the University of Pennsylvania mathematician Eugenio Calabi and the Chinese-American mathematician Shing-Tung Yau. These manifolds have a complicated shape that has the requisite geometrical properties for string theory. Researchers had initially hoped that there would be only one viable way to employ these extra six or seven dimensions, which would have resulted in a single version of string theory. But physicists were out of luck—there are multiple ways of accommodating the added dimensions.

Present incarnations of string theory have a 10- or 11-dimensional space-time, but this does not mean that the objects in string theory have these dimensions. Strings are one-dimensional objects within this space-time. However, despite the name, string theory includes objects other than just strings. In general, objects in string theory are called *branes,* a term derived from the word *membrane.* A one-dimensional brane, or one-brane, is a string. Higher dimensions can be incorporated into the theory, such as two-branes, three-branes, and so on. The world that people perceive could be construed as a three-brane embedded in a richer universe.

SUPERSTRINGS AND M-THEORY

The requirement for added dimensions is associated with an idea in physics known as *supersymmetry.* Supersymmetry pertains to the notion of balance in particle physics. In this idea, all particles have a partner, called a superpartner. No one has yet seen any of these hypothetical superpartners because they probably require more energy than particle accelerators can presently achieve, but if supersymmetry is true, then they exist. The pairing of particles is similar to the concept of antiparticles, as described in chapter 2, except supersymmetry involves the pairing of particles based on spin.

Pairing of particles in supersymmetry makes some of the advanced mathematics in theoretical physics easier. It also makes string theory a viable construct. The strings in supersymmetry string theory are called *superstrings,* which exist in a space-time having 10 or 11 dimensions.

Although supersymmetry makes string theory work, physicists do not know if supersymmetry is an accurate description of nature. In his 2002 article in *Astronomy,* Witten wrote, "Supersymmetric theories make detailed predictions about how superpartners will behave. To confirm supersymmetry, scientists would like to produce and study the new supersymmetric particles. The crucial step is building a particle accelerator that achieves high enough energies." The Large Hadron Collider (LHC), finished in 2008 and discussed in chapter 2, might be powerful enough to do the job. Witten added, "If supersymmetry can be confirmed in nature, this will begin the process of incorporating quantum mechanical ideas into our description of space-time." Or, in other words, combining quantum mechanics with general relativity.

Witten has made numerous contributions to string theory. Five different superstring versions of string theory, each with 10-dimensional

Institute for Advanced Study

Louis Bamberger, a New Jersey businessman and philanthropist, and his sister Caroline Bamberger Fuld, founded the Institute for Advanced Study at Princeton, New Jersey, in 1930. The first director, Abraham Flexner (1866–1959), had helped to make medical education much more rigorous in the early 20th century and brought considerable expertise to the institute. Goals of the institute did not include rapid growth and the addition of many faculty members but instead to provide a few gifted scientists with a home to come up with original thoughts and ideas. Originality and creativity are particularly important in science—if advanced problems could be easily solved with existing methods, someone would have already found the solution.

The list of past members of the Institute for Advanced Study is impressive. In the 1930s, shortly after the institute was established, Adolph Hitler came to power in Germany, and his fanatical ideas and hatred drove many talented scientists out of the country. One of these scientists was Einstein, who at the time held positions at the Kaiser Wilhelm Physical Institute and the University of Berlin. In 1932, Einstein left the country and settled at Princeton in 1933, taking a post at the Institute for Advanced Study, where he had presented several guest lectures earlier. Other well-known researchers who have worked at the institute include the computer pioneer John von Neumann (1903–57), and the physicist J. Robert Oppenheimer (1904–67), who led the scientific team that developed the atomic bomb during World War II.

Today the permanent faculty numbers about 40 (including emeriti [retirees]). But every year about 200 researchers from all over the world receive invitations to visit the institute and enjoy the academic freedom to pursue whatever scientific goals they think are most worthwhile. The Institute for Advanced Study has no formal links with other institutions or universities, but the proximity to Princeton University leads to a lot of collaborations.

space-time, had been worked out when Witten showed in 1995 that these versions are special cases of a more general theory. This general theory, which Witten called M-theory, involves tremendously complicated mathematics. (Witten has remarked on several occasions, such as in a 1998 article in *Notices of the American Mathematical Society,* that "*M* stands for magic, mystery, or matrix, according to taste.") M-theory needs a total of 11 dimensions—10 spatial dimensions and one dimension for time. Witten's employer, the Institute for Advanced Study, has been the cauldron for many profound ideas in science, including M-theory. The sidebar on page 153 provides more information on this noted institute.

Physicists do not know if M-theory is the most fruitful approach to take. But if it eventually leads to the theory of everything, perhaps the *M* will come to stand for "mother of all theories."

When deciding which, if any, version of string theory to pursue, theoretical physicists often rely on their mathematical and scientific instincts. But deciding if string theory, in any of its incarnations, is actually true is a matter for experiments and observations. And here lies some of the gravest difficulties facing string theory. In addition to the complicated mathematics, string theory does not make any easily verifiable predictions. To search for verification, researchers have been forced to think long and hard about the problem.

One possibility is to discover evidence for the extra dimensions required in all the current versions of string theory. A news release at Science*Daily* on February 4, 2008, sketches one idea from researchers at the University of Wisconsin-Madison and the University of California, Berkeley, led by the physicist Gary Shiu at Wisconsin. "There are myriad possibilities for the shapes of the extra dimensions out there. It would be useful to know a way to distinguish one from another and perhaps use experimental data to narrow down the set of possibilities," said Shiu.

Shiu and his colleagues wondered if any particle accelerator experiments would be able to discern the shape of additional dimensions, if they exist. The high-speed collisions in particle accelerators generate transient particles that quickly decay according to certain patterns. Specific properties of these particles determine the patterns seen in the accelerator detector, and Shiu and his colleagues found that, in theory, the shape of the extra dimensions would influence the behavior of hypothetical particles known as Kaluza-Klein gravitons. Shiu noted, "At least in principle, one may be able to use experimental data to test and constrain the geometry of our universe." The researchers proposed a set of experiments that may eventually be performed at LHC.

EXPERIMENTAL TESTS

String theory should make verifiable predictions, otherwise its usefulness will be limited. A theory that makes no predictions and offers no guidance for future experiments would be like a sports handicapper who only gave odds after the race was over. An accurate description of what has already happened makes a good historical record, but most scientists expect more. The Austrian-British philosopher Sir Karl Popper (1902–94) distinguished scientific theories from nonscientific theories on the basis of falsifiability—theories in science must be amenable to tests that could potentially prove the theory false. Scientific theories must go out on a limb, so to speak.

Verifiable predictions have been critical for successful theories in physics. Eddington's adventurous excursion, which tested one of the main predictions of Einstein's general relativity theory, is a prime example, and has become famous in the annals of physics. When Eddington confirmed Einstein's calculation on the bending of light, people took general relativity much more seriously. Experimenters have also subjected quantum mechanics to numerous tests, and the theory has passed every one.

The mathematical advances of the past few decades have made physicists realize that string theory might be able to unite quantum mechanics and general relativity. Perhaps string theory could even be a theory of everything. But before skeptics will accept the theory, it must be tested, meaning that it must generate verifiable—and falsifiable—predictions. In Cho's 2004 article in *Science,* the physicist Stuart Raby spoke for most scientists when he noted, "You're not going to believe string theory until you see the real world coming out of it."

A direct test of string theory would involve the existence of its namesake—strings. Proof that strings exist would demonstrate that string theory is grounded in reality. But while an observation of strings would be a momentous step, it is not likely to occur any time soon. Cho noted that "directly observing the putative strings would require collisions more than a million billion times more energetic than any that have been produced in a particle collider." And observation of strings may be out of the question on theoretical grounds. Recall that if strings are to be a theory of quantum gravity, the length of a string should be somewhere in the vicinity of Planck length. As discussed in the sidebar on page 146, probing objects at this scale requires a concentration of too much energy in a small space, potentially resulting in a black hole. Direct evidence for strings might never be possible.

If physicists can never see strings in a direct manner, they must come up with a more clever way of testing string theory. But scientists must be careful that these tests actually do what they are supposed to do.

For example, consider supersymmetry string theory, or superstrings, for short. Supersymmetry predicts that particles are paired with superpartners, which have not yet been found. If a researcher found evidence for one of these superpartners in, say, an experiment with a powerful particle accelerator such as LHC, the finding would strongly support supersymmetry. But this finding would not confirm string theory. The theory of superstrings needs supersymmetry in order to work, but supersymmetry does not depend on string theory. If superstrings are real, then supersymmetry is real also, but the opposite is not true—the existence of supersymmetry does not imply the truth of string theory. Supersymmetry could be real but string theory might not.

The absence of superpartners is also not strong evidence that string theory is false. Lacking firm calculations of the energy needed to generate these particles, a researcher could always suppose that the required energy is beyond the means of presently available particle accelerators.

Some people are skeptical that string theorists will be successful in their search for verifiable tests. String theory is mathematically rich and complicated, which means the theory is capable of explaining a lot of physics—perhaps too much. Along with the richness comes a certain amount of ambiguity, and an attempt at such generality that "covers all the bases" might lead to a lack of specificity.

Peter Woit, a researcher at Columbia University in New York, published a book in 2006, *Not Even Wrong: The Failure of String Theory and the Search for Unity in Physical Law*. Woit argues that string theory has dominated theoretical physics for the last few decades but has yet to produce any concrete advances because it has not been verified experimentally.

The failure to deliver verifiable predictions, or falsifiability as Popper would say, does not give string theory skeptics any reason for optimism.

In the worst-case scenario, strings may be a fascinating mathematical playground for gifted mathematicians to explore but have nothing to do with the way nature works. No one will know the truth until the theory is tested. While some theoretical physicists may decide to invest time on developing alternatives to string theory, other scientists are continuing to expend a lot of effort on discovering tests for string theory that can be done in practice as well as in principle.

SEARCHING FOR THE TRUTH

The completion of LHC, as discussed in chapter 2, was important in particle physics for many reasons, including the possibility of shedding light on string theory. In addition to searching for superpartners, LHC could play several roles in the further development, or maybe falsification, of string theory. A team of researchers from the University of California, San Diego, Carnegie Mellon University, and the University of Texas at Austin announced one possibility in 2007.

Particle interactions can be complicated, especially at high energies, but physicists have learned much by studying how particles behave during collisions or scattering experiments, such as the experiments that revealed the presence of quarks in protons. Future experiments may be able to test string theory in some fundamental way. Ira Rothstein, a physicist at Carnegie Mellon University, and his colleagues analyzed the scattering of particles known as W bosons; these heavy particles carry the weak force, and their discovery in 1983 at a CERN accelerator marked one of the triumphs of particle physics. LHC will be used to study these particles in more detail.

Rothstein and his coworkers found that W boson scattering experiments may provide a means of potentially falsifying string theory, if it fails the test. The University of California, San Diego, issued a news release on January 23, 2007, that quotes the UC San Diego physicist Benjamin Grinstein, who participated in the study: "The canonical forms of string theory include three mathematical assumptions—Lorentz invariance (the laws of physics are the same for all uniformly moving observers), analyticity (a smoothness criteria for the scattering of high-energy particles after a collision) and unitarity (all probabilities always add up to one). Our test sets bounds on these assumptions."

Jacques Distler, a physicist at the University of Texas at Austin who also participated in the research, noted that, "If the bounds are satisfied, we would still not know that string theory is correct. But, if the bounds are violated, we would know that string theory, as it is currently understood, could not be correct. At the very least, the theory would have to be reshaped in a highly nontrivial way."

The W boson test cannot confirm string theory, but it can falsify the theory, which means that string theory has finally ventured partway out on a limb. If the results of the test do not conform to what is expected from string theory, the theory is in trouble.

But LHC is going to be a busy place for years to come, and there is no telling when researchers will obtain enough data to make a determination on this issue. And, like any complex piece of equipment, LHC suffers from downtime—damage sustained while in operation on September 19, 2008, sidelined the accelerator for months.

Instead of devising methods to test the various aspects of string theory, some researchers believe that they can validate the theory by tying it to other theories that have already found broad support. Science-*Daily* posted a news release, "Physicists Connect String Theory with Established Physics," in May 3, 2007, describing a connection between strings and a type of theory called gauge theory.

Gauge theories play vital roles in many branches of physics, including quantum mechanics. The "gauge" in the theory is somewhat like a coordinate system, and the theory offers a description of a certain set of phenomena, such as particle interactions, on the basis of operations or transformations within this system. The standard model is an example of a specific type of gauge theory.

Researchers led by Igor Klebanov, a physicist at Princeton University, showed that some of the aspects of string theory link with gauge theory. In particular, Klebanov and his colleagues examined the behavior of quarks. Under normal conditions, the strong nuclear force binds quarks tightly into particles such as protons and neutrons. At high energies, such as achieved in powerful particle accelerators, quarks loosen up a little bit, allowing physicists to study them and model their behavior with gauge theory.

But when quarks are tightly bound, physicists have trouble understanding their behavior. Building on earlier work, Klebanov's team found that string theory can provide some insight into the behavior when the interactions intensify. In the news release, Klebanov gave an analogy: "It was as though our understanding was a road that started at the point where the interaction between quarks was weak. We could follow it for a few miles through greater and greater interaction strengths, but then it stopped before reaching the great strengths that exist in the atoms of rocks and trees—the section of road that string theory describes." Klebanov added that "there is hope that other facets of gauge theory are amenable to similar treatment." Niklas Beisert, a physics professor at Princeton University, said, "All these studies now make us sure that string theory and well-established gauge theory are indeed two sides of the same coin."

Even with these results, string theory still faces an uphill climb. Connections with particle physics and theoretical tests with particle accelerators are steps in the right direction, but some physicists are looking skyward for additional inspiration.

THE UNIVERSE WITH STRINGS

Since string theory is potentially a theory of everything, it can reasonably be expected to have cosmological implications. Perhaps scientists can find evidence for string theory in some kind of astronomical observation.

Researchers led by Mark Hindmarsh, a physicist at the University of Sussex in the United Kingdom, have begun to examine data from a National Aeronautics and Space Administration (NASA) satellite. This satellite, launched on June 30, 2001, contains the Wilkinson Microwave Anisotropy Probe (WMAP), which takes precise measurements of the cosmic microwave background radiation. The cosmic background radiation is a weak but steady radiation of microwave frequency that comes from all directions—it forms a background to astronomers who study the sky. Scientists believe that this radiation is the remnant or afterglow of the big bang, the explosion that created the universe some 14 billion years ago. The cosmic microwave background radiation is the oldest "light" of the cosmos. WMAP measures this radiation and detects any anisotropy, or variations, that gives clues to the geometry of the universe and how it evolved. The name of the probe honors David Wilkinson (1935–2002), a Princeton University cosmologist who was one of the earliest scientists to study the microwave background radiation.

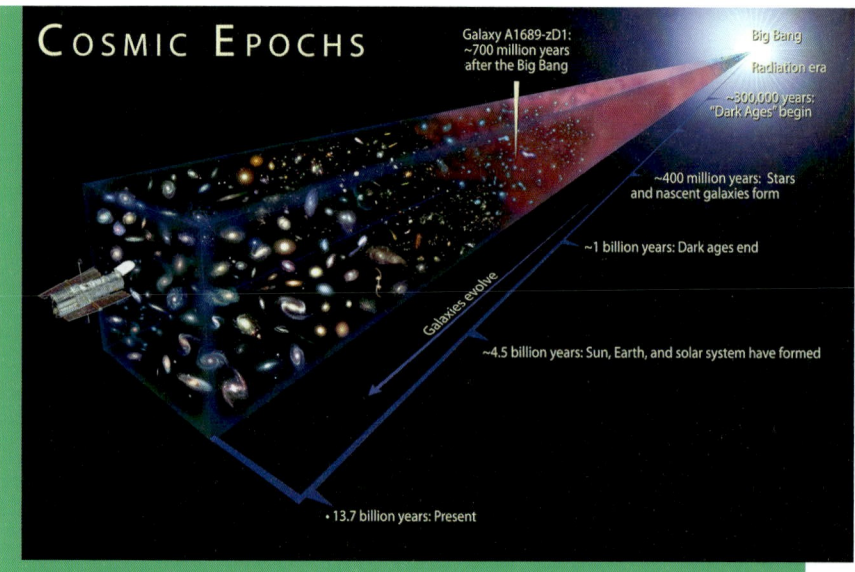

The history of the universe, from the big bang to the present state, is captured in this concise time line. *(NASA, ESA, and A. Feild [STScI])*

Hindmarsh and his colleagues studied the complicated mathematics of string theory to find some hints of how strings may affect the cosmic microwave background radiation. They used an extremely fast computer, called a supercomputer, to perform the complex calculations. Then they started to analyze WMAP measurements. Science*Daily* posted a news release, "Could the Universe Be Tied up with Cosmic String?" on January 21, 2008, in which the researchers announced preliminary findings. Although much more study is needed, the researchers found evidence supporting the existence of strings in the universe—"cosmic" strings. This evidence is indirect, but intriguing. Hindmarsh said, "This is an exciting result for physicists. Cosmic strings are relics of the very early Universe and signposts that would help construct a theory of all forces and particles."

Another team of researchers is looking at "old light" for different clues. As described in a news release that the University of Illinois at Urbana-Champaign released on January 28, 2008, a team of Illinois researchers had attempted to devise a test of string theory involving hydrogen absorption.

Atoms emit and absorb electromagnetic radiation at specific frequencies. For example, when a beam of radiation strikes a cloud of gas, the atoms in the gas absorb certain frequencies of radiation corresponding to the discrete levels of energy of their electron orbits, as given by quantum mechanics. After absorbing this energy, atoms tend to quickly emit it by sending out radiation of the same frequency. These interactions with light or radiation form a spectrum by which researchers can identify specific atoms from the observed frequencies of radiation. Scientists identify atoms by studying emitted light or light that has passed through a sample. An absorption spectrum contains dark lines at the frequencies where the atoms of the sample have absorbed light; atoms tend to emit this light quickly but do so in all directions, thereby scattering much of the light and removing most of it from the light beam.

Hydrogen is the most common element in the universe, so its spectrum is prominent. One of the most important spectral lines of neutral hydrogen—hydrogen that is not charged, or ionized, by the removal of its electron—occurs at a frequency of 1,420 million hertz (cycles per second). This frequency is in the microwave range. In a vacuum, this frequency of radiation has a wavelength of 8.4 inches (21 cm), which scientists often refer to as the "21 centimeter line." Astronomers focus on this line because it can penetrate the clouds of dust that hide many parts of the galaxy—this dust obscures radiation at many other frequencies.

According to string theory, strings would generate specific variations in the density of hydrogen gas in the universe. These characteristic variations should be detectable from the absorption spectrum of hydrogen in the early universe, which astronomers can observe by studying remote parts of the universe. But the expansion of the universe causes this radiation to be shifted toward the red end of the spectrum, toward lower frequencies and longer wavelengths. (The same shift occurs for all radiation, including the cosmic microwave background radiation.) As a result, the 21 centimeter line has been red-shifted by a factor of about 100, corresponding to radio frequencies.

Benjamin Wandelt, a researcher at the University of Illinois at Urbana-Champaign, and his colleagues found a method of detecting strings in the perturbations of this spectral line. But while this test is important in principle, the variations are so small that their detection would require a huge and costly array of radio telescopes, much more extensive than even the Very Large Array, a radio telescope array near Socorro, New Mexico, that stretches 22.3 miles (36 km) across.

CONCLUSION

At this point, string theory is the fruit of the labor of mathematicians and theoretical physicists. The promise of a theory of quantum gravity, or perhaps even a theory of everything, encourages this work. Yet after several decades, string theory has not moved off the drawing board, so to speak. It exists in elegant mathematical formulations rather than in experimental data.

Some physicists are discouraged. The title of Woit's book, *Not Even Wrong,* even suggests a sense of despair. If string theory cannot be brought into the laboratory at some point in its development—if its falsifiable tests are few and difficult or nearly impossible to conduct—then it will fail as a scientific theory. This is true even if string theory is correct, for if scientists cannot test the theory rigorously and thoroughly, most will refuse to put any faith in it.

But cosmological implications of string theory offer a glimmer of hope. The universe provides plenty of opportunities for observation and measurement.

Researchers at the University of Washington have investigated the possibility of using gravitation to test string theory. The university issued a news release, "Superstrings Could Add Gravitational Cacophony to Universe's Chorus," on January 8, 2007, announcing their results. Craig Hogan, a cosmologist at the University of Washington, believes that gravitational waves may hold the key to string theory.

Moving objects emit gravitational waves according to general relativity. A gravitational wave is a vibration in the curvature of space-time, as described by Einstein's general theory of relativity. Physicists have a lot of confidence in general relativity, and there is circumstantial evidence that gravitational waves exist, although no instrument is sensitive enough to measure these waves directly. Strings, if they exist, can also emit gravitational waves. Hogan has proposed that these waves, which will have low frequencies, could be an accessible means of finally detecting a string.

Attempts to measure gravitational waves have not succeeded because most waves are tiny, and noise from other sources tends to drown out the signal. But NASA has plans for a project called Laser Interferometer Space Antenna (LISA) that might succeed. An interferometer is a device that uses the interference of light waves to make precise measurements of distances. LISA will consist of three spacecraft positioned in an equilateral triangle, 3,100,000 miles (5,000,000 km) per side. Each of the spacecraft will carry sensitive instruments to keep track of any variations in the distance between it and the others. As gravitational waves ripple through this triangle in space, the distances will fluctuate, and LISA should be able to detect them.

The goals of LISA are to measure and study gravitational waves from many different sources in the universe, which will give astronomers and physicists an excellent tool to study the cosmos. Massive objects and violent collisions will form much of the data, but LISA may also provide ample opportunity to detect strings in the background, if Hogan is correct. The news release quoted him as saying, "If we see some of this background, we will have real physical evidence that these strings exist."

Plans call for a launch of LISA spacecraft in 2020. NASA has many other projects on its plate, so LISA could easily be delayed or canceled. But if and when it launches, physicists may get their first glimpse of strings.

Squeezing a rigorous test out of string theory has proven difficult, but some physicists are continuing to try to salvage this theory that promises so much. Witten wrote in his 2002 article, "String theory involves a conceptual jump that's large even compared with previous revolutions in physics. And there's no telling when humans will succeed in crossing the chasm." People may never succeed, yet they continue to try. Only by diligent effort and the determination to pursue promising leads whenever they arise will physicists at the frontiers of science arrive at the truth of the foundations of physics.

CHRONOLOGY

1687   The British physicist Sir Isaac Newton (1642–1727) publishes *Philosophiæ Naturalis Principia Mathematic* (*Mathematical Principles of Natural Philosophy*), containing his law of universal gravitation, among other important advances in physics and mathematics.

1900	The German physicist Max Planck (1858–1947) initiates the development of quantum mechanics with his proposal that energy is quantized.
1905	The German-American physicist Albert Einstein (1879–1955) publishes his theory on the quantum nature of light.
1916	Einstein publishes the general theory of relativity.
1926	The Austrian physicist Erwin Schrödinger (1887–1961) discovers an equation that becomes generally known as Schrödinger's equation, one of the foundations of quantum mechanics.
1927	The German physicist Werner Heisenberg (1901–76) proposes the uncertainty principle—measurements of certain variables, such as a particle's position and momentum, cannot be precisely obtained at the same time.
1934	The Russian physicist Dmitrii Blokhintsev and colleagues hypothesize the existence of gravitons.
1954	The Chinese-American physicist Chen-Ning Yang and the American physicist Robert Mills propose a type of gauge theory that becomes important in particle physics.
1974	A California Institute of Technology researcher John H. Schwarz and his colleague, Joel Scherk, discover a version of string theory that shows a lot of promise in unifying quantum mechanics and gravity.
1983	The Particle physicist Carlo Rubbia and his colleagues find the W and Z bosons.
1984	Schwarz and Michael Green find a mathematically rigorous way of formulating the equations of string theory.

1995	The University of California, Santa Barbara, researcher Joseph Polchinski identifies multidimensional objects in string theory known as D-branes.
	Edward Witten, a physicist at the Institute for Advanced Study, works out a general theory for strings called M-theory.
2007	The University of Washington physicist Craig Hogan and his colleague describe a string detector based on gravitational waves.
	The Princeton University physicist Igor Klebanov and his colleagues demonstrate some important links between string theory and gauge theory.
2008	Mark Hindmarsh, a physicist at the University of Sussex, and his colleagues announce that some data in their study of cosmic microwave background radiation indirectly suggests the existence of strings.

FURTHER RESOURCES

Print and Internet

Abbott, Edwin A. *Flatland: A Romance of Many Dimensions.* Mineola, N.Y.: Dover Publications, 1992. This book is a reprint of the 1884 original.

Cho, Adrian. "String Theory Gets Real—Sort of." *Science* 306 (11/26/04): 1,460–1,462. This news article reports on a conference on string theory held in Aspen, Colorado, in 2004.

Greene, Brian. *The Elegant Universe: Superstrings, Hidden Dimensions, and the Quest for the Ultimate Theory.* New York: W. W. Norton & Company, 1999. Greene, a physicist who has worked on string theory, provides an accessible account of the theory and its implications for the foundations of physics.

Kirkland, Kyle. *Particles and the Universe.* New York: Facts On File, 2007. Quantum mechanics, particle physics, and the theory of relativity can be difficult to understand, but this book presents an elementary discussion of the core concepts.

Public Broadcasting Service (PBS). "The Elegant Universe." Available online. URL: http://www.pbs.org/wgbh/nova/elegant/. Accessed June 22, 2009. This is the online companion of a PBS science program based on Brian Greene's book. Information on string theory and interviews with prominent scientists are included.

Schwarz, Patricia. "The Official String Theory Web Site." Available online. URL: http://www.superstringtheory.com/. Accessed June 22, 2009. Perhaps the term *official* in the title is a bit of a stretch, but this excellent Web resource has information on the fundamentals, mathematics, and history of string theory.

Science*Daily*. "Could the Universe Be Tied up with Cosmic String?" News release (1/21/08). Available online. URL: http://www.science daily.com/releases/2008/01/080120182315.htm. Accessed June 22, 2009. A team of researchers led by Mark Hindmarsh of the University of Sussex announces preliminary results suggesting that cosmic microwave background radiation may have patterns that support string theory.

———. "Particle Accelerator May Reveal Shape of Alternate Dimensions." News release (2/4/08). Available online. URL: http://www.science daily.com/releases/2008/01/080131161812.htm. Accessed June 22, 2009. Gary Shiu, a physicist at the University of Wisconsin-Madison, and his colleagues describe a series of accelerator experiments that may offer clues on the shape of the extra dimensions required in string theory.

———. "Physicists Connect String Theory with Established Physics." News release (5/3/07). Available online. URL: http://www.science daily.com/releases/2007/05/070502153818.htm. Accessed June 22, 2009. Igor Klebanov and his team at Princeton University discover a relationship between string theory and gauge theory.

Smolin, Lee. *The Trouble with Physics: The Rise of String Theory, the Fall of a Science, and What Comes Next.* New York: Houghton Mifflin Company, 2006. A skeptical physicist takes a hard look at string theory.

University of California, San Diego. "Physicists Develop Test for 'String Theory'." News release (1/23/07). Available online. URL: http://

ucsdnews.ucsd.edu/newsrel/science/stringtheory07.asp. Accessed June 22, 2009. Researchers Ira Rothstein, Benjamin Grinstein, Jacques Distler, and Rafael Porto announce that scattering experiments involving W bosons could lead to a test of string theory.

University of Illinois at Urbana-Champaign. "Scientists Propose Test of String Theory Based on Neutral Hydrogen Absorption." News release (1/28/08). Available online. URL: http://news.illinois.edu/NEWS/08/0128string.html. Accessed June 22, 2009. University of Illinois researchers develop a test for string theory involving astronomical observations.

University of Washington. "Superstrings Could Add Gravitational Cacophony to Universe's Chorus." News release (1/8/07). Available online. URL: http://uwnews.org/article.asp?articleid=29374. Accessed June 22, 2009. University of Washington scientists describe a method of using gravitational waves to detect strings.

Witten, Edward. "Magic, Mystery, and Matrix." *Notices of the American Mathematical Society* 45 (October 1998): 1,124–1,129. Witten discusses string theory and its relation to quantum mechanics and gravity.

———. "Universe on a String." *Astronomy* 30 (June 2002): 42–47. This accessible article on string theory describes some of the astronomical implications of the theory.

Woit, Peter. *Not Even Wrong: The Failure of String Theory and the Search for Unity in Physical Law.* New York: Basic Books, 2006. Woit, a mathematician and theoretical physicist, argues that string theory has failed to live up to its billing as a theory of everything.

Web Sites

Institute for Advanced Study. Available online. URL: http://www.ias.edu/. Accessed June 22, 2009. This Web site contains news and information on the Institute for Advanced Study in Princeton, New Jersey.

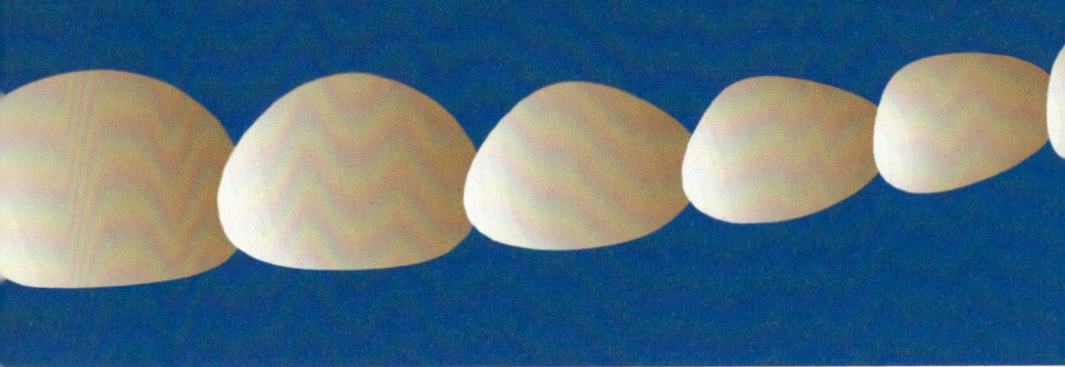

FINAL THOUGHTS

Despite the progress physicists have made since Newton formulated the universal law of gravitation and the laws of motion in the 17th century, the topics discussed in this book provide a sample of what is left to do. There is plenty to keep the next generation of physicists busy.

One of the many topics that merit a chapter in this book, if there had been sufficient room, is dark energy. The main reason it was not included is that scientists presently seem to have less of a handle on dark energy than other objects and phenomena at the frontiers of physics. Scientific understanding of dark energy is subject to substantial revision, if only because the current understanding is so rudimentary.

Dark energy is a mysterious type of energy that physicists hypothesized in 1998 after studying the expansion of the universe. Cosmologists believe that about 14 billion years ago a tremendous explosion called the big bang created the universe, which erupted from an incredibly hot, dense point and began expanding. The energy of explosions tends to dissipate over time; on Earth, the expanding gases and ejected objects slow down as they encounter air resistance, and in the universe, the force of gravitation exerts attractive forces on matter to counter the expansion.

But scientists who analyzed the spectra of a certain type of supernova made a startling finding. Supernovas of this type, called type Ia, share a lot of properties, and all of them have the same luminosity—they emit the same amount of light. At equal distances they would all have the same brightness, but as viewed from Earth, type Ia supernovas vary in brightness because they vary in distance, with the more distant objects being fainter. A brightness measurement of a type Ia therefore reveals its distance.

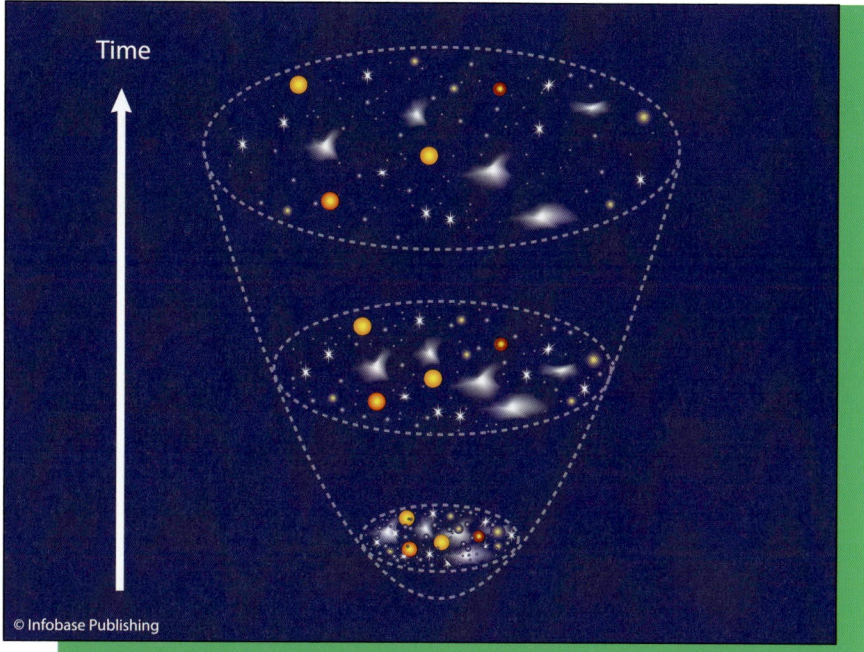

Time

© Infobase Publishing

This simple diagram illustrates the expansion of the universe from the big bang to the present time. As the universe expands, objects grow farther apart.

Astronomers can also examine a distant object's spectrum—the range of frequencies of electromagnetic radiation. Atoms and molecules emit or absorb light at specific frequencies, which make recognizable lines at these frequencies when astronomers look at the spectrum. But the universe is expanding, as illustrated in the figure above, and distant objects are moving away from each other. Because the frequency of radiation coming from a receding object is decreased, a phenomenon known as the Doppler effect, spectral lines of these objects are shifted downward, toward the lower or red end (so named because red light is the lowest frequency of visible light). This shift is called the redshift. A measure of redshift indicates the rate an object is receding from Earth.

The distance from a type Ia supernova, given by a brightness measurement, tells how long ago this short-lived phenomenon occurred, which equals the time required for light to travel this distance. Astronomers combined the distance data with a redshift measurement, which revealed how much the universe expanded since the supernova event took place. When scientists collected data from supernovas at various distances, they were surprised to find that the rate of expansion has been speeding up, or accelerating.

To explain this counterintuitive result, researchers hypothesized the existence of a mysterious type of energy that causes the expansion to accelerate by exerting a force that opposes gravitational attraction. This energy is called dark energy because its nature is obscure.

Subsequent observations have reinforced the supernova finding. Measurements of the cosmic microwave background radiation, which as described in chapter 6 is the remnant of the big bang, indicate that the universe is remarkably flat and uniform. Cosmologists have not found enough matter to achieve this uniformity, so they have hypothesized that dark energy accounts for the discrepancy. About three-quarters of the mass-energy of the universe may be dark energy.

Now that researchers have been alerted to the hypothetical existence of dark energy, they can search for other effects of this strange form of repulsive gravity. For example, scientists have used special telescopes to study the distribution of certain objects such as masses of hot gas. Hot or energized objects frequently emit high-energy radiation such as X-rays; Earth's atmosphere blocks most of this radiation—which is fortunate, otherwise people would be exposed to a dangerous dose of X-rays—so astronomers must study it with observatories positioned above the atmosphere. The Chandra X-ray Observatory is an orbiting satellite capable of detecting this form of radiation and transmitting the data to astronomers at the surface. The observatory, named for the Indian-American astrophysicist Subrahmanyan Chandrasekhar, was launched in 1999.

Using the Chandra X-ray Observatory, researchers studied masses of hot gas in clusters of galaxies, some relatively nearby and some remote. They found that the mass congregates in these clusters, as if the repulsive force of dark energy, which is causing space to expand, is controlling the growth and dispersal of these masses.

But scientists have no firm ideas about the nature of dark energy and the stuff of which it is made. Until researchers find a way of analyzing dark energy rather than just observing its effects, many plausible solutions to the puzzle can be entertained.

One controversial solution is that the puzzle is not a puzzle at all. For instance, consider the accelerated rate of expansion. The existence of some form of repulsive energy is one explanation, but other explanations, although equally mysterious, might apply. In 2008, the researchers Timothy Clifton, Pedro G. Ferreira, and Kate Land at Oxford University in the United Kingdom considered the possibility that Earth resides in a region of abnormally low density. This abnormality could be responsible for the observations relating to dark energy. As these researchers fully realize, a belief that Earth is in a "special" position, such as the center of the universe, has been associated with unscientific ideas in the past. Yet it could potentially explain the mystery as aptly as dark energy, although not without sacrificing a widespread assumption in science that Earth occupies no special place in the universe (except in the hearts and minds of its inhabitants).

Perhaps the real answer has yet to be grasped. Explaining dark energy, or its perceived effects, may require substantial revision in the way physicists think. But this is true for all the frontiers of science discussed in this book. The range and breadth of physics are vast, from hypothetical strings, atomic nuclei, subatomic particles, and neutrinos on up to superconductors, planetwide dynamical systems, and the expanding universe. All of these frontiers have opened up rewarding opportunities for the advancement of science, technology, and society.

APPENDIX A

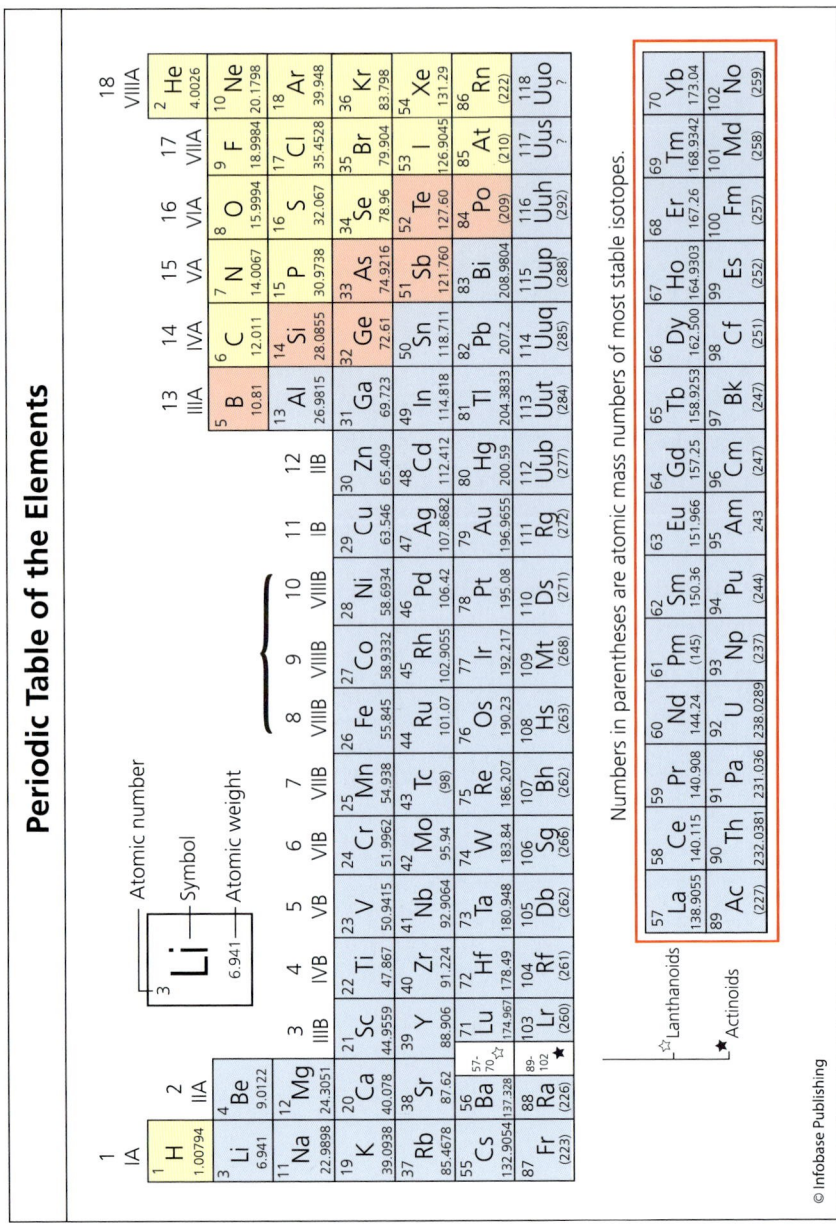

Periodic Table of the Elements

Atomic number

Symbol

Atomic weight

1 IA			
1 H 1.00794			
3 Li 6.941	2 IIA		
11 Na 22.9898	12 Mg 24.3051		
19 K 39.0938	20 Ca 40.078	3 IIIB	21 Sc 44.9559

3 Li 6.941

Numbers in parentheses are atomic mass numbers of most stable isotopes.

☆ Lanthanoids
★ Actinoids

© Infobase Publishing

172

APPENDIX B

The Chemical Elements

(g) none (c) nonmetallics

element	symbol	a.n.
carbon	C	6
hydrogen	H	1

(g) chalcogen (c) nonmetallics

element	symbol	a.n.
oxygen	O	8
polonium	Po	84
selenium	Se	34
sulfur	S	16
tellurium	Te	52
ununhexium	Uuh	116

(g) alkali metal (c) metallics

element	symbol	a.n.
cesium	Cs	55
francium	Fr	87
lithium	Li	3
potassium	K	19
rubidium	Rb	37
sodium	Na	11

(g) alkaline earth metal (c) metallics

element	symbol	a.n.
barium	Ba	56
beryllium	Be	4
calcium	Ca	20
magnesium	Mg	12
radium	Ra	88
strontium	Sr	38

(g) none (c) metallics

element	symbol	a.n.
aluminum	Al	13
bohrium	Bh	107
cadmium	Cd	48
chromium	Cr	24
cobalt	Co	27
copper	Cu***	29
darmstadtium	Ds	110
dubnium	Db	105
gallium	Ga	31
gold	Au***	79
hafnium	Hf	72
hassium	Hs	108
indium	In	49
iridium	Ir****	77
iron	Fe	26
lawrencium	Lr	103
lead	Pb	82
lutetium	Lu	71
manganese	Mn	25
meitnerium	Mt	109
mercury	Hg	80
molybdenum	Mo	42
nickel	Ni	28
niobium	Nb	41
osmium	Os****	76
palladium	Pd****	46
platinum	Pt****	78
rhenium	Re	75
rodium	Rh****	45
roentgenium	Rg	111
ruthenium	Ru****	44
rutherfordium	Rf	104

(g) none (c) metallics

element	symbol	a.n.
scandium	Sc	21
seaborgium	Sg	106
silver	Ag***	47
tantalum	Ta	73
technetium	Tc	43
thallium	Tl	81
titanium	Ti	22
tin	Sn	50
tungsten	W	74
ununbium	Uub	112
ununtrium	Uut	113
ununquadium	Uuq	114
vanadium	V	23
yttrium	Y	39
zinc	Zn	30
zirconium	Zr	40

(g) pnictogen (c) metallics

element	symbol	a.n.
arsenic	As*	33
antimony	Sb*	51
bismuth	Bi	83
nitrogen	N	7
phosophorus	P**	15
ununpentium	Uup	115

(g) none (c) semimetallics

element	symbol	a.n.
boron	B	5
germanium	Ge	32
silicon	Si	14

(g) actinoid (c) metallics

element	symbol	a.n.
actinium	Ac	89
americium	Am	95
berkelium	Bk	97
californium	Cf	98
curium	Cm	96
einsteinium	Es	99
fermium	Fm	100
mendelevium	Md	101
neptunium	Np	93
nobelium	No	102
plutonium	Pu	94
protactinium	Pa	91
thorium	Th	90
uranium	U	92

(g) halogens (c) nonmetallics

element	symbol	a.n.
astatine	At*	85
bromine	Br	35
chlorine	Cl	17
fluorine	F	9
iodine	I	53
ununseptium	Uus*	117

(g) lanthanoid (c) metallics

element	symbol	a.n.
cerium	Ce	58
dysprosium	Dy	66
erbium	Er	68
europium	Eu	63
gadolinium	Gd	64
holmium	Ho	67
lanthanum	La	57
neodymium	Nd	60
praseodymium	Pr	59
promethium	Pm	61
samarium	Sm	62
terbium	Tb	65
thulium	Tm	69
ytterbium	Yb	70

(g) noble gases (c) nonmetallics

element	symbol	a.n.
argon	Ar	18
helium	He	2
krypton	Kr	36
neon	Ne	10
radon	Rn	86
xenon	Xe	54
ununoctium	Uuo	118

* = semimetallics (c)
** = nonmetallics (c)
*** = coinage metal (g)
**** = precious metal (g)

a.n. = atomic number
(g) = group
(c) = classification

© Infobase Publishing

GLOSSARY

absolute zero the coldest possible temperature, which equals -459.67°F (-273.15°C), and is represented in the Kelvin scale as 0 K

antimatter material that behaves similarly to ordinary matter but with the opposite or reverse of certain properties such as charge

antiparticle antimatter partner of a particle

attractor in dynamical systems, a point or region to which the system tends to converge

beta decay a kind of radioactive event in which an electron and a type of neutrino is emitted

black hole an object with a gravitational field so powerful that nothing, not even light, can escape it

bosons particles that have certain spin values and can mediate or carry the forces of nature

branes string theory objects that may have various spatial dimensions

cold fusion an event or process in which nuclei fuse without need of high temperature

conductor a material that permits the flow of electricity

Cooper pair in superconductor theory, two electrons that become linked and move through materials without resistance

cosmic rays   high-energy particles and radiation that enter Earth's atmosphere from space

critical temperature   in superconductors, the temperature at which the material becomes superconducting

cyclotron   an accelerator that propels and guides ions in a circular path with electric and magnetic fields

dark energy   a hypothetical substance that is responsible for accelerating the expansion of the universe

dark matter   substances that are obscure or invisible, but have mass and exert gravitational forces

deuterium   an isotope of hydrogen that has one neutron

dimensions   aspects, measurements, or variables that characterize a system or object

dynamical system   a set of variables whose state evolves, as registered in phase space

electron neutrino   a lepton whose interactions involve another fundamental particle, the electron

electron volt (eV)   unit of energy in particle physics equal to the kinetic energy gained by an electron as it accelerates through a potential difference of one volt

eV   See **electron volt**

fractal   object that displays self-similarity, so that each part has a structure similar to the whole

general relativity   See **general theory of relativity**

general theory of relativity   description of the gravitational force as a curvature of space-time

greenhouse gases   substances that cause warming by trapping heat

hadrons   class of particles that consist of quarks and can experience strong nuclear forces

initial conditions   the state or value of variables at the beginning point in time

ions   charged particles

isotopes   members of an element that differ in the number of neutrons in the nucleus

Kelvin scale   a measurement of temperature in which the unit has the same magnitude as the Celsius degree but absolute zero is represented with the value 0 K

leptons   fundamental particles that have half-units of spin and are not subject to the strong nuclear force

muon neutrino   a lepton whose interactions involve another fundamental particle, the muon

neurons   brain cells whose electrical activity governs thought and behavior

nuclear reactors   means of producing energy from reactions involving atomic nuclei

nucleons   general name for the protons and neutrons composing the nucleus

nucleus   the central part of an atom, containing positively charged protons and electrically neutral neutrons packed tightly together

phase space   a mathematical construction that consists of the set of all possible states of a system

phonons   vibrations in a solid

plasma   a state of matter consisting of ions in the gaseous state

quantum gravity   quantization of gravitational fields by applying the principles of quantum mechanics

quantum mechanics   a set of principles and equations that predict the behavior of particles in terms of probability

quarks   a family of fundamental particles that interact so strongly that they are not found free in nature, but instead compose hadrons such as protons and neutrons

radioactive exhibiting the property of certain atomic nuclei to emit high-energy radiation

resistance in electricity, opposition to the flow of current

space-time the four-dimensional construct of the three dimensions of space and one dimension of time

spin in quantum mechanics, a characteristic property of particles that tends to obey the same laws of angular momentum

strong force See **strong nuclear force**

strong nuclear force the strongest of the four fundamental interactions, mediated by particles called gluons and important in the structure of the nucleus

supernova an exploding star

superstrings objects in versions of string theory that incorporate supersymmetry

supersymmetry a hypothetical concept in which particles are paired with partners called superpartners

tau neutrino a lepton whose interactions involve another fundamental particle, the tau

tritium a radioactive isotope of hydrogen that has two neutrons

turbulence irregular or disorderly pattern of motion

wavelength the distance between crests of a wave or vibration, consisting of one full cycle

weak force See **weak nuclear force**

weak nuclear force a fundamental interaction, somewhat weaker than the strong nuclear force, mediated by W and Z bosons

X-rays high-energy electromagnetic radiation with a frequency much higher than visible light

FURTHER RESOURCES

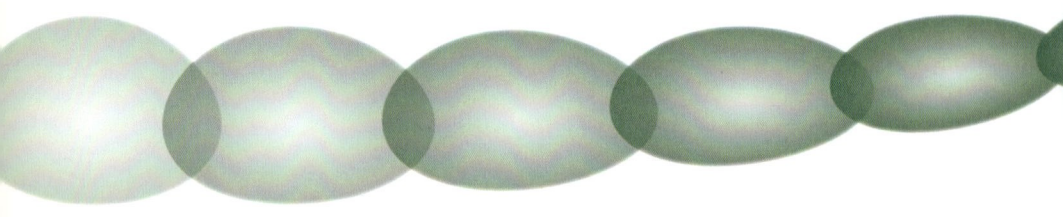

Print and Internet

American Physical Society. "Physics Central." Available online. URL: http://www.physicscentral.com/. Accessed June 22, 2009. These Web pages contain articles and images that explain the science of physics and its applications. Included are introductions to some of the physicists who are expanding the frontiers of science.

Bloomfield, Louis A. *How Things Work: The Physics of Everyday Life,* 4th ed. New York: Wiley, 2009. This textbook offers an excellent introduction to physics by explaining the concepts in terms of familiar objects.

Calle, Carlos I. *Superstrings and Other Things: A Guide to Physics.* Oxford, U.K.: Taylor & Francis, 2001. Calle explains the laws of physics in a clear and accessible manner.

Cropper, William H. *Great Physicists: The Life and Times of Leading Physicists from Galileo to Hawking.* Oxford: Oxford University Press, 2001. Twenty-nine chapters relate the biographies of Albert Einstein, Max Planck, Wolfgang Pauli, Werner Heisenberg, Marie Curie, Ernest Rutherford, Enrico Fermi, Murray Gell-Mann, and others.

Einstein, Albert. *The Special and General Theory.* New York: Penguin Classics, 2006. This book is a reprint and translation of a short book that Einstein wrote to explain his theories to the general reader.

Feynman, Richard P., Robert B. Leighton, and Matthew Sands. *The Feynman Lectures on Physics. I. Mainly Mechanics, Radiation, and Heat.* Boston: Addison Wesley, 1963. This is the first volume of a set of books based on the late Professor Feynman's inimitable lectures on physics.

The book is suitable for beginning college students and advanced high school students.

————. *The Feynman Lectures on Physics. II. Electromagnetism and Matter.* Boston: Addison Wesley, 1964. This is the second volume of a set of books based on the late Professor Feynman's inimitable lectures on physics. The book is suitable for beginning college students and advanced high school students.

————. *The Feynman Lectures on Physics. III. Quantum Mechanics.* Boston: Addison Wesley, 1964. This is the third volume of a set of books based on the late Professor Feynman's inimitable lectures on physics. The book is suitable for beginning college students and advanced high school students.

Gonick, Larry, and Art Huffman. *The Cartoon Guide to Physics.* New York: HarperCollins, 1991. Gonick's cartoon guides explain the basic concepts of a subject with the aid of cartoons. This book is an entertaining but serious look at the core principles of physics.

Holton, Gerald, and Stephen G. Brush. *Physics, the Human Adventure: From Copernicus to Einstein and Beyond,* 3rd ed. Piscataway, N.J.: Rutgers University Press, 2001. This book offers a comprehensive review of the important problems and discoveries in the history of physics.

Lederman, Leon M., and Christopher T. Hill. *Symmetry and the Beautiful Universe.* Amherst, N.Y.: Prometheus Books, 2004. Symmetry is the ability to stay the same after some sort of change—for example, an equilateral triangle looks the same after it has been rotated 120 degrees. This concept plays an important role in particle physics and other topics in physics, as explained in this book.

Nave, Carl R. "HyperPhysics." Available online. URL: http://hyper physics.phy-astr.gsu.edu/hbase/hframe.html. Accessed June 22, 2009. Hosted by the Georgia State University Department of Physics and Astronomy, this educational resource has sections on mechanics, electricity and magnetism, light and vision, relativity, condensed matter, nuclear physics, quantum physics, heat and thermodynamics, sound and hearing, and astrophysics.

Nicolson, Iain. *Dark Side of the Universe: Dark Matter, Dark Energy, and the Fate of the Cosmos.* Baltimore, Md.: Johns Hopkins Press, 2007.

This book provides an introduction to the latest concepts of matter, energy, and the evolution of the universe.

Nobelprize.org. "The Nobel Prize in Physics." Available online. URL: http://nobelprize.org/nobel_prizes/physics/. Accessed June 22, 2009. The Nobel Foundation has issued a prize in physics since 1901. This Web resource contains the list of winners, their biographies, and the lectures they delivered at the presentation ceremony.

Plait, Philip C. *Bad Astronomy*. New York: Wiley, 2002. This book discusses some common mistakes in physics and astronomy and explains why they are wrong.

Web Sites

Exploratorium. Available online. URL: http://www.exploratorium.edu/. Accessed June 22, 2009. The Exploratorium, a museum of science, art and human perception in San Francisco, has a fantastic Web site full of virtual exhibits, articles, and animations, including much of interest to physicists and physicists-to-be.

How Stuff Works. Available online. URL: http://www.howstuffworks.com/. Accessed June 22, 2009. This Web site hosts a huge number of articles on all aspects of technology and science, including physics.

National Aeronautics and Space Administration (NASA). Available online. URL: http://www.nasa.gov. Accessed June 22, 2009. NASA's Web site contains a huge amount of information on astronomy, physics, and Earth science, and includes news and videos of NASA's many exciting projects.

Science*Daily*. Available online. URL: http://www.sciencedaily.com/. Accessed June 22, 2009. An excellent source for the latest research news, Science*Daily* posts hundreds of articles on all aspects of science. The articles are usually taken from press releases issued by the researcher's institution or by the journal that published the research. Main categories include matter and energy, space and time, earth and climate, and others.

INDEX

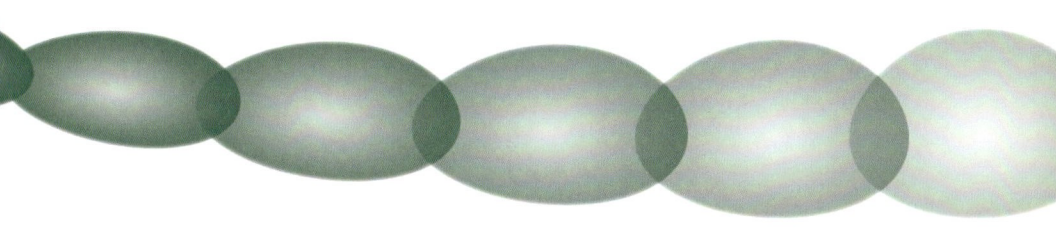